MEN

MEN

EVOLUTIONARY AND

LIFE HISTORY

RICHARD G. BRIBIESCAS

HARVARD UNIVERSITY PRESS

Cambridge, Massachusetts, and London, England

Copyright 2006 by the President and Fellows of Harvard College

ALL RIGHTS RESERVED

Printed in the United States of America

First Harvard University Press paperback edition, 2008

Library of Congress Cataloging-in-Publication Data
Bribiescas, Richard G.
Men: evolutionary and life history / Richard G. Bribiescas.
p.   cm.
Includes bibliographical references and index.
ISBN 978-0-674-02293-5 (cloth : alk. paper)
ISBN 978-0-674-03034-3 (pbk.)
1. Men.   2. Men—Physiology.   3. Human evolution.   I. Title.
HQ1090.B743 2006
306.4—dc22   2006049542

To Pop

CONTENTS

MEN

Chachugi's Cap

IT DOESN'T TAKE A DEGREE in biology to notice that men and women are utterly different. Physical and behavioral contrasts pervade everyday life. Men, on average, are taller and physically stronger than women. Men don't have babies. Men make up about half of the human population, but they are involved in far more than half of humankind's destructive and violent behavior. Men age differently than women do, and men die sooner and at a more rapid rate than women.

Male and female humans also differ in significant ways from the point of view of those who study human evolution. Human reproductive success, a central theme in evolutionary biology, is much more difficult to trace for men than it is for women. Many of the differences are obvious. Young men do not experience a defining event such as the onset of menstruation (menarche), so it is not easy to identify just when they become sexually mature. Unlike women with their cyclical ovulation, men produce gametes (sperm) continuously throughout their reproductive years. And men's reproductive years last longer than women's: for men there is no acute depletion of gametes that is analogous to female menopause. Furthermore, while it is usually clear which woman is an infant's mother, paternity is difficult to determine with comfortable certainty, thereby making assessments of male reproductive success very problematic. For these reasons, far more work has been done on female reproductive biology than on that of males.

The central topic of this book is not solely whether men are different—from women or from nonhuman males—but also why and how. What are the physiological and evolutionary causes of the differences? How similar are male humans to males of other species? Are they saddled with the evolutionary baggage of countless previous generations of mammals? Such questions require an approach that addresses both mechanistic and evolutionary causes: a sense of the historical factors that shaped men's evolution as well as an understanding of the physiology that was the focus of natural selection.

In contemporary biology, evolutionary theory and life history theory are the dominant guiding principles that motivate research. Briefly, evolutionary theory explains the origins and development of species through time, while life history theory provides an explanation of the evolution of important life events such as growth and reproduction in a species. Both these sets of ideas are powerful intellectual tools that have been shown to be valid and accurate by innumerable experiments and years of research. As will become apparent in later chapters, in many ways, evolutionary theory and life history theory are intertwined; it is difficult to invoke the ideas of one without those of the other. Life history theory is not a new mode of thought, but it has only recently been recognized as an important tool for understanding human evolution and found its way into the discipline of biological anthropology. As a guiding theory, it is a living set of ideas that change with the incorporation of new data.

Since life history theory focuses on the timing of and the investment of energy in particular aspects of physiology, such as growth and reproduction, biological anthropologists have focused on these two areas in their efforts to understand human life histories. Given the central role of reproduction in natural selection and evolution, reproductive function has become an important concern of life history research. This particular field has been dubbed reproductive ecology, in reference to the interaction between the physiology of an organism and its response to environmental influences.

Anthropologists embraced life history theory decades ago in their demographic analysis of human populations. Nancy Howell's book on the Dobe !Kung of southern Africa opened the door for demographic research in anthropology (Howell 1979), and many other anthropologists were also using demography to answer key questions related to tribal peoples and to the evolution of human life histories. However, very few in those early days conducted physiological research with life history theory as a guide. Reproductive ecologists in several laboratories were among the first to examine human physiology through a life history lens and to employ field methods that allowed quantitative measurements of physiological variables such as hormone function in remote populations (Campbell 1994; Ellison 1988; Worthman and Stallings 1994). Such work made it clear that the methods and theoretical guidance were available to address questions about the evolutionary physiology of human males.

This is not say that human male physiology was a new field of research. Traditionally, it has been the domain of medical investigators. Clinical researchers have made important contributions to our understanding of male physiology, while evolutionary biologists have conducted groundbreaking work on quantifying and testing aspects of life history and evolutionary theory in nonhuman organisms. More recently, biological anthropologists have begun to play a central role in melding these two perspectives, combining the theoretical guidance of evolutionary and life history theory with quantitative methods previously reserved for medical researchers.

In my first season of field work as a biological anthropologist, I traveled to Paraguay to work among the Ache, indigenous South American people who until recent years had subsisted as nomadic hunter-gatherers, and who had come into peaceful contact with the outside world in the 1960s and 1970s. The evolutionary physiology of human males was to be the focus of my research, but my overall strategy was not yet clear. What was clear was that the study of males required a different theoretical and analytical approach from the study of females.

Some of the techniques developed for studying women could be used with men, however. Researchers had learned a great deal about the relationship between women's reproductive function and the stresses women faced from their environment by collecting samples of saliva and analyzing them for levels of reproductive hormones. Such work had been done among people of the Ituri Forest in Zaire (now the Democratic Republic of the Congo), as well as in Nepal (Ellison et al. 1989b; Panter-Brick et al. 1993). I hoped to build on this foundation, applying the techniques of salivary steroid analysis to Ache males.

The challenge for me was to develop a new strategic angle in order to have an intellectual tool that would produce testable hypotheses. Determining the feasibility of testing these hypotheses experimentally in the field would be the next step. Before forming a question I had to find out if Ache testosterone levels were different from those of American men.

My initial research plan, therefore, was quite simple. I hoped to collect samples of saliva from Ache men and assess salivary testosterone in order to get some sense of the variation in testosterone levels. I had no idea how the men would react to my request for saliva. To my great relief, they were willing to cooperate, and in fact they seemed amused by my interest in their spit. They nicknamed me Emberygi, which, roughly translated, means "spit guy."

One morning a colleague and I crossed paths with an Ache man by the name of Chachugi. Like many of the Ache in recent years, Chachugi wore donated clothes handed out by missionaries. He was wearing a baseball cap, the kind you can buy in any convenience store in the United States. The English words printed on the front caused me to pause. The bold black letters on a white background said this:

There are three stages to a man's life: Stud, Dud, Thud.

I asked Chachugi if he minded if I took his photograph. He didn't mind: you can see him in Figure 1.

Throughout my time in Paraguay and later, I thought about the words on Chachugi's cap. They seemed to be an apt summary of the life history of male humans. Indeed, they described the life history of the males of many species.

Many aspects of Ache life fascinated me. Although I was a newcomer and had only a rudimentary understanding of the Ache language, I could increasingly notice which men were influential in community life and which ones were more peripheral. Some men opted out of the male role altogether. *Panegis,* from an Ache word meaning "unlucky at the hunt," were men who did not participate in the male world. They were

Figure 1 Chachugi.

not homosexual, or at least not according to the Aches' description, but men who simply chose to assume a female role in society, including all the chores and tasks that were commonly performed by women, such as cooking and childcare. Panegis were often not the most kindly treated individuals, but they had an accepted place in the community.

Toward the end of my field season, I stopped at a hut on the periphery of an Ache village. There, a panegi around the age of sixty—I'll call him Japegi—lived with his extended family. I was there to pick up my last batch of saliva samples and to thank the family members for their help. In return for the samples, they asked for some salt, sugar, and other items, exchanges to which I gladly agreed. Then, after some whispered discussions, they told me that Japegi also had a request. This was unusual; during my previous visits to their hut, Japegi had merely smiled and never said anything to me. He had also struck an unusual pose whenever I arrived, kneeling, facing the fire, almost as if he was trying to hide his face. From my lofty position as anthropologist, I assumed that his body language reflected the demure attitude expected of a panegi. However, today he spoke. In halting, broken Spanish, he asked for *calzoncios*—underwear. The true reason he had been kneeling during my visits was not some cultural rule of body posture. His demeanor simply reflected embarrassment that his poverty had left him naked from the waist down.

This was an important lesson for me as an anthropologist. It brought home the danger of basing interpretations on preconceptions. In science we search for parsimonious explanations: that is, when faced with a question we are supposed to consider basic answers first, before weighing more abstract or esoteric alternatives. In my speculation about Japegi's behavior, my jump from assumption to anthropological fact was clearly in error. In this case, perhaps, a simple query to his family would have been useful: ethnography has an important place in anthropology. However, many questions cannot be answered by ethnographic inquiry; instead, they need to be formulated in a fashion that produces testable hypotheses. Explanations and answers that are based solely on subjective interpretation or are otherwise untestable are often counterproductive and do not serve to support anthropology as a science.

If the words on Chachugi's cap have stayed with me as a concise description of male life history, the life of a panegi illustrates that male life histories, whether in western industrialized societies or in remote areas of the world, are complex, involving physiological and behavioral aspects about which scientists have much to learn. Over the years since my encounters with Chachugi and Japegi, I have become convinced that the combination of biological research with history theory will help us address many questions that remain unanswered about the evolution of human males. Important strides have been made toward understanding female reproductive function from an evolutionary and life history perspective. Although the methods and theoretical development of male reproductive ecology will follow a different path from the study of females, the basic underpinnings of evolution will still apply. In the following chapters I will set out what these perspectives have enabled us to learn, to date, about the evolutionary and life history of male humans.

PART I

Setting the Stage

I

Change Happens

MALE HUMANS HAVE A BIOLOGICAL history. Businessmen commuting on a Tokyo subway, olive growers pruning their trees on a Spanish hillside, and hunters stalking a capuchin monkey through a forest in Paraguay are all descendants of hominid ancestors that probably emerged on the African continent about two hundred thousand years ago. Moreover, despite their technological and cultural differences, none of these groups of men can be considered more evolutionarily advanced or primitive than the others. As organisms, all men are the consequence of millions of years of evolution, and all use their inherited physiology to cope with their respective environments.

The extent to which evolutionary processes affect the present state of human existence is a topic of debate. Some scholars argue that cultural and social influences now far outweigh any effects of natural selection on contemporary human physiology (Sahlins 1976). Others suggest that culture is subject to Darwinian forms of evolution and acts in conjunction with biological evolution (Richerson and Boyd 2005). I don't think any biological anthropologist would deny that cultural and social forces have a tremendous effect on how humans, including males, behave and how their physiologies respond to environmental challenges. However, humans carry evolutionary baggage. Physiology has been shaped by past selection pressures, most likely by pressures very similar to the ones that pervade the everyday life of contemporary hunter-gatherer populations. It would be naive to imagine that human populations are not currently undergoing differential fertility and mortality. And those processes are at the essence of natural selection, the shaping mechanism of evolution.

Certain criticisms surrounding the use of evolutionary theory to understand contemporary human populations are, however, valid. Shoddy and irresponsible research tainted by political agendas has made its way into the academic mainstream (for discussions see Gould 1981; Marks 1995; Shipman 1994). But to totally discount evolutionary theory because of its previous misuse would be to outlaw matches in response to the crimes of arsonists. Many criticisms of the investigation of contemporary human populations within an evolutionary context are rooted in the misconception that evolution implies determinism: that evolutionary theory corners our species into a cul-de-sac of limited potential. The age-old argument of nature versus nurture is misguided. An organism's physiology (nature) is in many ways meant to respond to a changing environment (nurture). An unyielding physiology is seldom advantageous since the environment is never static. In other words, are human males the product of nature or nurture? The answer is yes.

The following discussion of males, their physiology, their development throughout history, as well as their present state, requires some understanding of two central biological theories: evolution by natural selection and life history theory. In many ways these modes of thought are intertwined, but for the sake of clarity, I will treat them as distinct but complementary concepts for now.

The Big and Small Pictures

As an undergraduate I had the good fortune of working in the laboratory of the neurobiologist John Liebskind, who pioneered research on endogenous opiates, substances secreted by nerve cells that act as natural painkillers. At about the same time, I took a course in evolutionary anthropology from the biological anthropologist Nadine Peacock, who would play a pivotal role in my decision to pursue biological anthropology as a career. Professor Liebskind's research made me think about the process by which interesting questions for study were identified. How do you choose which protein to work on? Which neuron cluster should be the focus of the next experiment? Is this simply trial and error, or is there some way of developing a long-term strategy based on a well-defined and supported set of ideas? Physics has relativity, quantum mechanics, and principles of thermodynamics. What guides biology? Professor Peacock instilled the importance and the utility of evolutionary theory in spawning hypotheses within research on human biology.

But what sort of questions can evolutionary theory address? In general, two categories of evolutionary study, micro- and macroevolutionary processes, are commonly recognized. The difference between micro- and macroevolution is in scales of time. Macroevolutionary processes are what people commonly associate with evolutionary theory: the formation of species (speciation) and the taxonomic relationships between organisms. A basic macroevolutionary question concerns the identification and understanding of factors that cause speciation and delineate one species from another. How is a species defined? How are different species related (Mayr 2004)? Within anthropology, such questions are the primary domain of paleoanthropologists, who study the when, where, and how of human speciation and the relationships between the numerous species that have been associated with the human lineage. These are daunting tasks given the rarity of hominoid and anthropoid fossils and the stakes involved in tracing the evolutionary history of our own species.

Microevolutionary studies involve examining the effects of contemporary selection factors on living organisms in order to observe evolutionary processes occurring in real time, or at least within a relatively short period. To accomplish this task, biologists attempt to manipulate physiological and environmental variables that are potentially related to mechanisms that affect reproduction and survival. For example, within the field of reproductive ecology, investigations of variation in hormone levels in response to environmental challenges such as undernutrition and heavy manual workloads can provide insights into physiological mechanisms that control the allocation of energy to different bodily (somatic) needs. This strategy has been central to biological anthropologists' investigation of male reproductive function and life history. Demographic anthropology, meanwhile, has focused on quantifying the effects of natural selection by observing changes and patterns in fertility and mortality within a population. Changes in these parameters of life history are important reflections of evolution in action.

Nonrandom Elimination

The brilliant evolutionary biologist Ernst Mayr summarized the origins of the terms we use when discussing natural selection this way:

The difficulty begins with the exact description of the process of selection. After Darwin had discovered his new principle, he searched for an appropriate terminology and thought he had found it in selection, the term animal breeders used for the choice of their breeding stock. However, as first Herbert Spencer and then Alfred Russel Wallace pointed out to him, there is no agent in nature which, like the breeders, "selects the best." The beneficiaries of selection are the individuals that are left over after all the less fit individuals have been eliminated. Natural selection thus is a process of "nonrandom elimination." Spencer's statement, "survival of the fittest," was quite legitimate, provided the term fittest is properly defined. (Mayr 1997: 2091)

At its essence, selection is simply the removal of certain individuals and their genetic variants from a population, either through higher mortality or through decreased fertility. The process is "nonrandom" because the individuals who are eliminated possess certain characteristics that make them less likely to successfully produce offspring who will carry their genes into future generations. Those who are not eliminated are the "fittest"—that is, those who have higher fitness are those who make a greater genetic contribution to the next generation.

Imagine a world with absolutely no constraints on population growth. Houseflies, to take one example, produce thousands of offspring every year. If all were to survive, after only a few days the Earth would be a pulsating ball of fly biomass. The same would hold true for slowly re-producing organisms such as elephants. The time it would take for elephants to fill every square inch of real estate would obviously be greater, but it would eventually happen. The power of exponential growth was what led the English clergyman and economist Thomas Robert Malthus to predict that, unless family size was regulated, the human population would outpace the supply of resources such as food, resulting in shortages and competition (Malthus 1798).

Darwin found inspiration in this message when working on his theory of evolution by natural selection. He noted in his 1876 autobiography:

> In October 1838, that is, fifteen months after I had begun my systematic inquiry, I happened to read for amusement Malthus on *Population,* and being well prepared to appreciate the struggle for existence which everywhere goes on from long-continued observation of the habits of animals and plants, it at once struck me that under these circumstances favourable variations would tend to be preserved, and unfavourable ones to be destroyed. The result of this would be the formation of a new species. Here, then I had at last got a theory by which to work. (Darwin 1958, 120)

Darwin also noticed aspects of natural history and geology that provided him with key insights. He had seen fossil remains in South America that were similar but not identical to modern extant forms. This suggested to him the idea of extinction and of a nonstatic natural world. Other natural historians prior to Darwin had challenged the notion of a static world in which all of the species alive in modern times came into existence at the Earth's creation; now Darwin sensed that change was occurring over time. But how much time? Interpretations of the biblical account of creation had placed the age of the Earth at about six thousand years. Geologists such as Charles Lyell, in contrast, argued that if processes such as erosion had always occurred at the extremely slow rate that was being observed in his own day, the world must be much older than a few thousand years (Lyell 1842). Darwin saw change over time, a very long time.

But how did change occur? The question was one of mechanism. Enter Malthus. Darwin postulated that if resources were limited, as they often are, competition would arise between individuals. The ones with traits that enabled them to outcompete their rivals would thrive, and more important, would reproduce and pass along those favorable traits to their offspring. When Darwin was formulating his ideas, he was free from many of the issues that are debated today regarding the idea of adaptiveness. In simple terms, a trait that is adaptive is one that increases the fitness (reproductive success) of the individual. Darwin's basic idea of an adaptive trait involved characteristics like neck and leg length and shell shape among Galapagos tortoises. Tortoises that lived on drier islands had longer necks, longer legs, and carapaces (shells) that turned upward above the neck and legs, allowing them to reach vegetation in the taller bushes that were characteristic of drier islands. Tortoises that lived in moister environments, where the edible vegetation was closer to the ground, had shorter necks and legs and less upturned carapaces. Using these observations along with his knowledge of animal husbandry and selective breeding, Darwin came to the conclusion that several basic conditions were necessary for evolution by natural selection to occur.

First, there must be variation in a trait. For without variation there cannot be the possibility of alternative solutions to an environmental challenge. Second, these traits must be heritable. This was something of a leap of faith for Darwin since he did not know how traits were passed from generation to generation. It was not until the importance of Gregor Mendel's work was recognized in the twentieth century that the mechanics of heritability were hammered out and genetic processes were incorporated into evolutionary theory (see Mendel and Krizenecky 1965). Other concerns involving heritability were vexing problems for Darwin. Critics challenged him with several quandaries, including the maintenance of variability. If traits were heritable and subject to blending, as was commonly observed in selective breeding of animals and crops, how was trait variation maintained? In other words, if you mated a black-haired animal with a white-haired one, critics argued, you would eventually get gray and lose all variation. For Darwin this problem was troublesome throughout his life; it was eventually addressed with the incorporation of Mendelian genetics. As a final condition, there must be differences in reproductive success or fitness. It was not enough for an individual organism to survive, it had to produce more offspring than its rivals. (Today we know that it is more accurate to say that organisms evolve to produce the *optimal* number of offspring.)

In simplest terms, to evolve means to change. Change can occur quickly or slowly. Sometimes it appears as if change refuses to occur. Years ago I was watching a nature special on television when the narrator stated that the shark had not changed in millions of years—that its evolution had stopped. Nonsense. The small amount of change in basic shark morphology (form and structure) in fact meant that selection was getting rid of many of the variants. That is the only way for an organism to stay so morphologically fixed: through natural selection and the elimination of variants.

Natural selection is the mechanism that induces change in a population, sometimes resulting in a new species. But the term "natural" can be misleading. It can be taken to imply that selection guided by humans, as in animal and crop breeding, is not natural but artificial. The term "artificial" also implies that humans are detached from the natural world. Given the extraordinary impact human populations have had on the ecology of this planet, this is somewhat nonsensical. The dichotomy between "natural" and "artificial" selection is also deceptive because it suggests that evolutionary theory has a moral agenda of some sort, which it does not. In evolutionary theory, there is no good, bad, natural, or artificial, and humans are included as part of the living world. Darwinian evolutionary theory does not support any moral stance. When evolutionary biologists speak of the adaptiveness of infanticide, they are not condoning the killing of babies. They are simply trying to explain a phenomenon from the perspective of evolutionary theory. This is not to say, of course, that scientists are devoid of social responsibility in conducting research. But the moral or ethical implications of any biological research are matters that are inherently unrelated to evolutionary theory.

Ernst Mayr's statement that natural selection can be best described as "nonrandom elimination" is correct. For males, the process of nonrandom elimination involves, among other things, efforts to be preferred over other males by females, competition with other males over females themselves or over resources that are crucial to being preferred by females, competence in keeping offspring alive and mates fertile, and strategies to minimize the risk of raising another male's progeny. These factors are as central to male existence today as they were when the first hominids roamed the African landscape.

Favoring Bad Ideas?

Feral donkeys once roamed freely in and around the mountain community of Big Bear Lake, California, where my brother and his family live above Los Angeles. During a visit as we were finishing dinner and listening to the raucous brays of a lovesick stallion, my sister-in-law asked why they made such a ridiculous sound when trying to gain the attention of a mare. My response was that the female donkeys, for some reason, preferred males who made this vocalization, absurd as it might sound to human ears. Stallions who could not or would not bray in a fashion that attracted mares would not leave many progeny behind. Female choice is a powerful selective mechanism for male traits and one of the many reasons why physical traits and behaviors change over generations.

But how did this braying evolve? There must have been variation among males. Perhaps in some ancestral population, some males brayed and others did not. Or perhaps all brayed but the sounds they made differed in some manner that was distinguishable to females. Perhaps variation in braying reflected some real measure of male genetic quality that could be passed to offspring. Traits that were central to braying must have had some genetic roots and been capable of being passed to offspring. But a loud bray surely wasn't an entirely advantageous characteristic. There are other animals that still live in the mountains above Los Angeles: black bears, coyotes, and the occasional mountain lion. Braying may not always be a good idea if one needs to be inconspicuous so as not to end up on the menu of a predator.

The males of many species, including humans, exhibit traits that seem to be bad ideas. The overgrown claws of male fiddler crabs and the risk-taking behavior of the human male are prime examples. However, before we conclude that a trait is detrimental to survivorship, we need to show that the trait is associated with a cost such as increased morbidity or mortality. In humans, of course, experimenters cannot ethically modify traits in order to increase mortality. It is much easier to study such questions in other organisms. Fortunately for evolutionary biologists, certain conditions and constraints are observed in males of most organisms, so that conclusions are often applicable to humans.

It's time to introduce another type of selection: sexual selection. While natural selection is often the result of competition between individuals, sexual selection involves competition within each sex, competition to attract or get access to mates. The different reproductive agendas of males and females are also components of sexual selection. This selective process favors traits that enhance reproduction. Sexual selection illustrates a central life history trade-off in males that will be a theme in later chapters: the trade-off between energy devoted to securing opportunities to mate (reproductive effort) and energy devoted to staying alive (survivorship).

In fiddler crabs, the males' oversized claw appears to put them in danger from predators. One of their primary predators is the great-tailed grackle *(Quiscalus mexicanus)*. The biologist Tsunenori Koga and colleagues demonstrated that male fiddler crabs are subject to much higher predation rates than females by grackles that employ an angled mode of attack. The males' large claws and brighter coloration make them more conspicuous to birds that employ this tactic (Koga et al. 2001). And the oversized claw, though it looks formidable, is virtually useless as a means of protection.

So is there evidence that large male claws are the result of sexual selection? Do the claws increase males' opportunities to mate, either by giving them an advantage in competition with other males or through preferential choice by females (or both)? Evidence indeed suggests that females prefer larger claws and that the claws also serve to block the entrances of burrows where females lay their eggs. A male waves his hefty claw, attracts a female, and subsequently uses the claw to keep other males from entering the burrow. Large claws are also associated with enhanced fighting ability against other males. Interestingly, restriction of males' access to food decreases their rate of claw waving, a primary way of signaling claw size and male condition to females (Jennions and Backwell 1996, 1998). Underfed males wave less and attract fewer females, illustrating the relationship between male somatic condition and reproductive effort, a concept we will return to later.

The role of sexual selection in humans is not so clear-cut, but there is compelling evidence that female choice and competition between males have had a significant effect on the evolution of human males. Men today are, on average, larger than women, although fossil evidence indicates that this difference has been toned down compared with that in some of our hominid ancestors. Species characterized by such sexual dimorphism (disparity between the sexes) in size are also believed to exhibit intense male-male competition (Plavcan 2000). Demographic data show that male mortality as the result of risky behavior is associated with a period in a man's life when reproductive effort is likely to be at its peak. Finally, because of their physiology, men have shorter average life spans than women even when one controls for the greater male tendency to engage in risky behavior. Testosterone seems to be an important culprit in compromising the life spans of human males.

If Darwin's theory of evolution is correct, why should there be traits that undermine individuals' ability to survive? The answer lies in a subtle but important modification in evolutionary theory. Darwin had no knowledge of genetics. With the incorporation of genetics in the twentieth century, it became clear that the central unit of natural selection was not the body as a whole but the gene. As far as evolution was concerned, the body was merely a carrier for its genes, and reproductive success meant the survival and propagation of the genetic material, not necessarily that of the individual. Darwin knew that, somehow, reproductive success was key to understanding these apparent discrepancies, but without knowledge of genetics, he simply could not provide a mechanism.

The Modern Synthesis

Mendel's writings on genetics, published in the 1860s, remained unnoticed for many years. It wasn't until 1900 that several botanists independently rediscovered his work and recognized its importance.

In the 1930s, biologists also recognized the centrality of genetics to evolutionary research and unraveled the connections between genetics and Darwinian evolution. Through the development of several theories that conciliated statistical population genetics and Darwin's original dilemma of reconciling continuous variation with particulate forms of inheritance, the biologists and statisticians R. A. Fisher, Sewall Wright, and J. B. S. Haldane at last showed that Darwinian evolution by natural selection was indeed the primary mechanism of organismic development through time (Fisher 1930). Field investigations by the biologists Ernst Mayr, Theodosius Dobzhansky, and George Gaylord Simpson provided strong supporting evidence for their predictions, and evolutionary theory entered a new era of predictive power and growth (Dobzhansky 1970).

Geneticists such as Walter Sutton and Thomas Hunt Morgan were working out the mechanisms of inheritance. Sutton predicted that inheritance units were located on chromosomes, the small structures that were visible under a microscope in each cell nucleus. He noticed that chromosomes had existed in pairs, which split apart during meiosis, the cell division that results in the formation of a gamete. In a tiny laboratory at Columbia University, Morgan demonstrated that genes were indeed located on chromosomes through his investigation of traits in drosophila (fruit flies) (Morgan 1915).

The molecular structure of genes did not become known until 1953, when Francis Crick and James Watson, relying on work by Rosalind Franklin, demonstrated the double helix organization of deoxyribonucleic acid (DNA). Following this discovery, a very clear picture began to emerge regarding such processes as mutation, transcription, and translation, the mechanisms associated with generating genetic variation and transforming these genetic plans into physical realities. It had been theorized that random changes or errors in the replication of genes were the basis for formation of novel mutations that natural selection could act upon. However, the location and biochemical nature of these changes, until this breakthrough, had been poorly understood.

In reference to the evolution of male humans, these developments in evolutionary theory provided researchers with the tools to grapple with puzzling issues such as reconciling selection for traits that favored survivorship as predicted by Darwin and Wallace with the observation that many individuals, especially males, seemed dead set on either killing themselves or developing traits that were clear handicaps against staying healthy and sound. The incorporation of genes as units of selection into evolutionary theory and the importance of the soma as a carrier of genes, proved to be a crucial factor in explaining, what seemed at the time, to be inexplicable.

Relatives, Genes, and Diminishing Returns

Even after the modern synthesis, certain issues continued to plague evolutionary theory. One of these was the observation of behavior that

appeared contrary to Darwin's prediction that organisms should behave selfishly in order to survive and reproduce. Individuals were expected to look out for their own interests and generally compete with
one another—and yet members of a number of species exhibited behavior that appeared to be altruistic or self-sacrificing. Biologists had long
noted that bees sacrificed themselves for the sake of the hive. When a
bee stings you, it tears its stinger from its body and eviscerates itself in a
process that leaves it mangled and dying. In addition, most of these
eviscerated individuals never reproduced. If an altruistic trait is somehow embedded in the physiology of an individual, it is by definition influenced by genes that are passed through reproduction. Individuals
who developed a "niceness" mutation would ultimately be selected out
of the population since the emergence of a selfish mutant would reap
the benefits of the altruists while the selfish individual would gain all of
the benefits with little or no cost to itself. In essence, altruistic genes
would be on the fast track toward extinction. But the observation that
bees, wasps, ants, and even some mammals were altruistic toward individuals was initially baffling.

Initial attempts to explain altruism involved a concept known as
group selection. The premise was that certain traits of individuals within
a group might be selected for because they benefited the fitness of the
group as a whole, whether or not they benefited the individuals. For example, given limited environmental resources, self-restraint might be a
selectable trait: the group might be better off if some members limited
their reproduction to restrict population growth and avoid overburden-
ing the environment (Wynne-Edwards 1962). However, there are several
challenges to this concept. First, cheaters would have to be kept at bay.
The introduction of a mutation for selfishness would soon eliminate the
altruistic genes: as the selfish individuals produced more offspring than
their altruistic fellows, the trait of selfishness would proliferate in the

group's population. It is possible, although unlikely, that group selection may occur in nature, but only within specific conditions (see
Richerson and Boyd 2005).

Biologists such as W. D. Hamilton and J. B. S. Haldane challenged
the notion of group selection when they realized that a refinement of
Darwin's earlier definition of fitness was needed, one incorporating not
only individual reproductive success but the reproductive success of genetically related individuals, such as siblings and cousins. Inclusive
fitness, as this is known, is defined as an individual's own reproductive
success (fitness) plus any increased fitness of relatives that results from
that individual's actions. In other words, if you save the life of your sibling's child, it counts toward your inclusive fitness. If you had nothing
to do with the rescue, it doesn't count. Genes may be propagated
through the reproduction of individuals that share similar genes through
common descent. Haldane foreshadowed the development of this idea
when, according to popular biology lore, when asked if he would give
his life to save a drowning brother, he replied, "No, but I would to save
two brothers or eight cousins," implying that through common descent,
these relatives in total would pass along his genes.

In 1964, Hamilton expanded on this idea and produced a work that
introduced an entirely new approach to evolutionary theory. He noted
that organisms that exhibited a high frequency of altruistic behavior—
such as social insects like bees, in which most individuals are sterile and
devote their energy to supporting and defending the hive and allowing the queen to reproduce—were genetically different from mammals.
They were haplodiploid. Females were diploid, bearing genetic material
from both a mother and father. Males, though, were products of unfer-
tilized eggs with no genetic contribution from a father. Without expand-
ing on the genetics, suffice it to say that a typical hive is highly inbred,

and even though most of the females are sterile, their efforts to sustain the hive result in reproductive payoffs through the proliferation of alleles they share with other members of the hive (Hamilton 1964).

However, mammals and most vertebrates are not haplodiploid. Kin selection, as Hamilton's theory had been dubbed, did not address the apparently selfless behavior observed in birds and some mammals. Eventually, many altruistic behaviors, such as the alarm calls with which ground squirrels warn of the approach of a predator, would be understood through kin selection. Individuals performing this altruistic behavior did so only under specific circumstances in which the beneficiaries of the behavior were close relatives (Sherman 1981). Similar results were noted in many other species, including primates (Chapais et al. 2001).

Nonetheless, individuals of some species seemed to reproduce well below their maximal capability without any close relatives receiving any benefit. This observation appeared to be evidence supporting the concept of group selection. For example, great tit birds *(Parus major)* seemed to limit their own egg production. If natural selection favored maximal reproduction, this species was clearly an exception. Advocates of group selection stated that the birds were limiting their reproduction in order to avoid overpopulation. However, the ornithologist David Lack offered an alternative explanation based on an elegant series of experiments. He showed that *Parus* produced, on average, nine eggs. If researchers removed a newly laid egg from the nest, the female would quickly lay another one. As long as the eggs were removed, replacements would be produced—to the point where it was quite apparent that the females had the physiological capability to produce many more eggs than they usually did. By manipulating the number of eggs in a nest, Lack illustrated that simply producing eggs is not the same as reproductive success. He observed that there was an optimal number of eggs to produce the greatest number of offspring over the life span of the mother. (In basic terms, an individual's lifetime reproductive success is the number of reproductive years of an adult multiplied by the number of offspring produced annually discounted by the probability of each offspring surviving to reproduce; see Murray and Bertram 1992; Brown 1988.) Very large clutches suffered from very high mortality, primarily because the parents could not support so many offspring. Very small broods had much lower mortality but were limited by the initial number of eggs. The average clutch size of nine eggs was revealed to be the optimal size (Lack et al. 1957; see also Visser and Lessells 2001). Thus individual birds were not restraining their reproduction out of an altruistic urge to avoid overtaxing their environment for the benefit of the larger group, and their behavior did not support group selection after all.

Within the social sciences, group selection is still invoked as an explanation of altruistic behavior (Borrello 2005). This is not to say that Hamilton's theory can explain all human altruistic acts, but it is important to note that human behavior is subject to the same limitations and constraints faced by other organisms. In addition, the trade-off between number of offspring and their probability of surviving to reproduce will be an important topic when we address fatherhood. The most basic understanding of male reproductive strategies suggests that males should father as many offspring as possible, since their physiological investment in the reproductive process is minor compared with that of females. However, we will find that under many circumstances males have been selected to stay and aid in the rearing of offspring, even though in doing so they may be missing opportunities to father more progeny.

Male Adaptations

During a visit to the Maine shore, a friend's toddler son emerged from the surf in a panic, making a beeline toward his mom. When he reached her, he hooked his thumbs into the front of his swimming trunks, pulled mightily, and peered down at his body, screaming, "They're gone! They're gone!" The lad had been unaware of the effect cold ocean water would have on the micromusculature surrounding his scrotum. The production of sperm is a temperature-sensitive endeavor, and human males have evolved ways of keeping the testes neither too hot nor too cold.

Evolution by natural selection results in physical traits that aid in optimizing lifetime reproductive success. Simply stated, traits that evolved to aid in optimizing reproductive success are known as adaptations. Adaptation is a concept that is central to evolutionary theory but is often subject to considerable debate. It is easy to find oneself using circular definitions when trying to identify adaptations. This is not to say that evolutionary biologists doubt the existence of adaptations. It is the subtlety of defining and testing adaptations that usually sparks debate. However, the concept of adaptation is indispensable in the effort to identify, define, and quantify physical traits that regulate an organism's physiology in response to challenges such as predation, the need to forage efficiently, and the need to attract mates (Futuyma 1986). The evolution of male physiology can be seen as selection that favored the development of adaptations, that is, physical traits that foster optimal lifetime reproductive success.

Identifying adaptations is often an intellectually thorny task, but there are some ground rules that prevent overzealous evolutionary biologists from attaching the term "adaptive" to every speck of an organism's morphology simply because an adaptive tale seems to make sense (Fisher 1985). Human males are chock full of weird and interesting traits that beg for an evolutionary explanation, but providing such explanations requires careful consideration of what can be empirically tested as an adaptation.

The first challenge is to identify the candidate trait. Remember that evolution is the differential proliferation of alleles (genetic variants) in a population, so for purposes of testing it is desirable to identify a trait that is as close a representative of an allele or set of alleles as possible. Some traits are easily identifiable and can be associated with specific genes. For example, genes of a specific class, called homeotic or Hox genes, have an important integrative function in determining the overall physical organization of an organism. By altering the expression of a Hox gene, geneticists can create a fruit fly with a leg where its proboscis should be (Abzhanov et al. 2001). Humans also possess Hox genes, but similar alterations are quite rare and obviously cannot be induced experimentally (Goodman and Scambler 2001).

One strategy for identifying the adaptive significance of a chosen trait is to manipulate the trait and observe the effect on particular aspects of fertility or survivorship. For example, the production and effects of testosterone in men are regulated in part by several genes that code for proteins such as luteinizing hormone (LH) and its complementary receptors. Moreover, there are small but significant differences in LH protein structure that result from genetic variation (Haavisto et al. 1995; Huhtaniemi and Pettersson 1999). It is therefore reasonable to assume that testosterone physiology has been subject to natural selection. So what would happen if we manipulated testosterone levels? This sort of manipulation is commonly done in humans for a variety of clinical reasons, but because of the long human life span, observing changes in mortality or fertility over time is not easy. We can, however, manipulate testosterone levels in other organisms and observe these changes. Results from studies of animals have shown this to be a quite useful strategy for accumulating data supporting the adaptive utility and significance of a trait.

A second method of identifying the adaptive significance of a trait is less direct but more practical. One can define a trait, garner information on variation in that trait within a population, and observe differential fertility and mortality in association with the variation. But how do we define a trait?

A particularly useful strategy is to observe and compare aspects of life history that are common to all organisms: growth, maintenance, reproduction, and senescence. It is difficult to argue against the idea that selection has acted upon traits that are central to these basic processes of life. It does not matter if we are speaking about a hundred-year-old oak tree, a newborn guppy, or an adult human. Each one of these organisms has inherited a physiology that has evolved mechanisms—adaptive traits—to deal with these inherent needs. Examining these traits in male humans and comparing them with those observed in other organisms is a powerful approach to questions about the evolution and the adaptiveness of human male physiology.

2

Birth, Death, and Everything in Between

LIFE HISTORY THEORY is the total quantitative assessment of the life of an organism. Traditionally, life history theory is segregated into three general fields of interest: demography, quantitative genetics, and trade-offs. Each of these addresses different questions regarding the evolution of an organism.

Demographic analyses of age-specific fertility and mortality within a population are used to quantify and assess the effects of selection pressures. Demography, in its simplest sense, is counting people. Measuring the effects of natural selection on the evolution of a population involves observing fertility and mortality—the number and rate of births and deaths—particularly within certain age classes. For example, changes in mortality in women of reproductive age will have a greater impact on population growth than such changes in postmenopausal women. Anthropologists utilize demographic methods to quantify the effects of environmental factors on population dynamics, since while selection acts on individuals, populations are what evolve. Remember, natural selection is nonrandom elimination, so any change in population structure that cannot be attributed to random selection (also known as genetic drift) is potentially natural selection in action.

Quantitative genetics attempts to unravel the interactions of genes with other genes and with environmental influences. Some genes exhibit partial expression or are only fully expressed when another specific gene or set of genes has been activated. Adaptations that vary in response to changing environmental conditions are known as reaction norms. Why would there be selection for these reaction norms? Environments change. Therefore there has been intense selection for crucial physiological mechanisms to be malleable and responsive to these changes. With over three billion base pairs in the human genome, quantitative genetics is obviously an extremely complex discipline. It is also a central part of life history theory, since it addresses microevolutionary processes in which variations are expressed in response to environmental conditions.

Finally, trade-off analysis attempts to understand how organisms adjust to an ever-changing environment by allocating time and energy to various physiological needs (Stearns 1992). In essence, life history theory provides a suite of methods for studying processes that are common to all species, and therefore it makes it possible to formulate testable hypotheses about, for example, the timing and organization of crucial lifetime stages and events such as growth span, reproductive maturation, and death. We will look at life history trade-offs in more detail later in this chapter.

Comparative Research

Life history theory has only recently been applied to questions related to human evolution and physiology. However, evolutionary biologists have used the predictions and premises of life history theory to examine growth, development, reproduction, and senescence in a broad range of other organisms. Comparative research, the investigative methodology of testing a hypothesis by correlating the results of two or more different categories of organisms, such as species or orders, is an important aspect of life history research. Testing predictions of life history theory requires comparing different species in order to determine the universality of mechanisms associated with time and energy allocation (Harvey and Pagel 1991). It also aids in determining which traits are invariant and therefore not likely to be subject to natural selection (Charnov 1993). However, analytical comparisons must be done with caution. While some traits, such as the fins in dolphins and penguins, are results of convergence (the process by which similar selection pressures lead to the development of analogous traits) some others, such as the fins of salmon and sharks, are shared because two species have a recent common ancestor. Such distinctions are important because variable independence is crucial to conducting valid tests of a hypothesis.

The predisposition to exhibit certain traits because of their presence in an organism's evolutionary lineage is called phylogenetic inertia. To understand it a bit better, imagine that a prey animal, say a species of antelope, had wheels, eight-cylinder engines, and four-wheel drive instead of legs. These traits would surely be to the antelope's advantage when trying to outrace a hungry lion on the African savanna. And yet the likelihood of such automotive traits evolving in an antelope, a zebra, a human, or a warthog over the next few million years (or ever) is quite remote. This is not because such traits would not confer any advantage, but because the genes particular organisms inherit from their ancestors constrain the possibilities of evolving certain traits. For example, it is highly unlikely that humans will evolve indeterminate skeletal growth—

the ability to keep growing throughout their lifetimes—a trait that is exhibited by many reptiles and fish. The reason is the long period of evolutionary separation between reptiles and mammals and their independent evolution of genetic diversity.

How influential is phylogenetic inertia in contemporary human males? Boys and men do face many of the same reproductive and physical constraints as males of other mammals such as deer, lions, and chimpanzees. For example, internal gestation and internal fertilization, two traits that are common to mammals, were extremely important in shaping human male physiology and behavior. These traits have resulted in differences in the amount of energy mothers and fathers invest in offspring as well as in uncertainty about children's paternity. The significance of these mammalian characteristics in male humans will become apparent in later chapters.

Studies involving organisms as diverse as salmon, fruit flies, worms, and great apes have all contributed information that is useful in the attempt to understand human males. The comparative method is a powerful tool in evolutionary biology, and when used correctly it can provide invaluable insights into evolutionary and life history processes. There are solid reasons for studying other animals to learn about humans: we cannot experiment on human subjects in the same manner as we can on nonhuman organisms. Current standards of morality, regardless of whether we agree with these or not, allow scientists to manipulate and often sacrifice animals' lives in scientific experiments. Research on animals has clearly benefited humankind: without it we would not have been able to develop many of the drug therapies currently available to address human diseases. But there are limitations on what we can glean about humans from research on nonhuman animals.

All species have distinct life histories. Some are more similar than others and share important traits, often because they are descended from a relatively recent common ancestor. Herein lies a problem. Much of what we know about physiology has come from nonhuman organisms, mainly rodents and nonhuman primates. Many studies of primates have focused on Old World monkeys, usually rhesus macaques *(Macaca mulatta)*. Macaque and human life histories have diverged in several major ways, including the manner and timing of their maturation, reproduction, and senescence. Other primates, and even our closest evolutionary relative, the chimpanzee, also differ from humans in important characteristics, including age at first reproduction, body weight, and maximum life span. The life histories of rodents, with their characteristic traits of multiple offspring per litter, small body size, and short life span, are even less similar to those of humans.

Nevertheless, comparative data can add to our understanding of human male life histories. Humans do share many aspects of life history with nonhuman primates, rodents, and other members of the class Mammalia. Internal gestation by females, determinate growth, and homeothermy (the ability to generate and regulate one's own body temperature) all reflect distinct traits that humans share with other mammals and indeed with many other vertebrates. There are many significant differences, of course, that set male humans apart from other male mammals. However, the methodology of investigating and understanding human males involves the same theoretical tools that have been so useful in other species.

Allocating Time

All organisms face time constraints. The amount of time an organism has to live is dictated by several factors, including its rate of senescence (basically how fast it falls apart before it dies) and its ability to stave off extrinsic sources of mortality such as predators and infectious agents. These factors have resulted in a wide range of variation in the life spans of organisms, from several hours in some insects to hundreds or even thousands of years in the case of bristlecone pine trees. The rate at which an organism senesces is largely a function of the number of hazards in its environment. If you're a rabbit in a forest full of predators, it may be a wise idea to mature and reproduce quickly and abundantly since your chances of surviving from one day to the next are dubious at best. If you're an Atlantic clam resting quietly and safely in the muck several hundred feet from the ocean surface, you can take your time to grow as large as possible (or never cease to grow) and reproduce at a steadier rate (Austad 1997b).

The challenge of time allocation is very simple. No one can be in two places at once. In life history theory, the principle of allocation of time is a function of current and long-term evolutionary processes. The current evolutionary aspect involves behavioral strategies by which individuals attempt to budget daily activities efficiently. Organisms must decide when to forage for resources, search for mates, and care for offspring, to name just a few tasks. Time devoted to the pursuit of one goal is usually time that cannot be devoted to others, but the tasks sometimes overlap. Diversions from optimal foraging schedules, for example, can have potential influences on fertility and survivorship (Hirota and Obara 2000; Hurtado and Hill 1987). Indeed, the impact of foraging patterns and subsequent needs for energy has been used to model the behavior and ecology of ancient hominids (Sorensen and Leonard 2001). Although direct consequences for fitness are not easily evident in humans and primates, because of their long lives and complex behavioral strategies, ample evidence, especially from hunter-gatherer populations, suggests that human behavior is strongly affected by selection for an optimal balance of time allocated to various tasks (Cronk 1991a).

For males, a common trade-off in time allocation is between food-seeking and mate-seeking behaviors. Foraging for food enhances survivorship; energy devoted to finding mates results in declines in foraging efficiency. These behavioral shifts in time investment may lead to changes in body composition that compromise survivorship, such as the breakdown and use of fat and increases in metabolically expensive tissue such as skeletal muscle mass. This is evident in seasonally breeding mammals such as deer. When males enter the rut, they spend their time fighting other males, defending territory, and coercing females to mate with them. During this time, males forgo eating in favor of seeking opportunities to mate. This decision to allocate time to acquiring access to mates and not to eating has very direct consequences for the males' condition and mortality (Bobek et al. 1990). The parallel with male humans is not exact, but these observations do suggest that male physical investment in body mass, which has different effects on survivorship and reproductive effort, has probably been influenced by millions of years of time-allocation decisions.

Males of many species are more likely to be caught by predators during the mating season, because mating behavior makes them more conspicuous to predators or because they neglect to pay attention to their surroundings. One example of increased predation that is related to investment in mating is seen in South American frogs. During the mating season, male frogs come together to form choruses that call in an attempt to attract females. However, certain species of bats recognize these frog mating calls as signals of the location of a good meal, and inflict a significant toll on male frogs eager to find mates. Some large males croak loudly, both attracting many willing females and becoming preferred targets of predatory bats. However, the fitness benefits of attracting and mating with many females probably offset the costs to survivorship. Softer-croaking males do not attract many mates but are not as subject to predation. They seem to have "chosen" greater survivorship over short-term gains in access to mates (Ryan 1980; Tuttle and Ryan 1981).

Another aspect of males' time trade-offs is mate guarding. In organisms with internal fertilization, including humans, uncertainty about paternity is a key factor in male reproductive strategies. Not wanting to invest time and energy in progeny that are not genetically his, a male may apply anticuckoldry tactics such as guarding his mate, vigorously attempting to restrict other males' access to her. This is a taxing proposition for several reasons. First, females have their own reproductive and life history agendas and will often choose to mate with other males given the opportunity. Second, males will sometimes form coalitions to oust a male who has sequestered one or more females. Finally, in some species, some males have evolved behavioral and physical characteristics that deflect the guarding male's attention away from them so they can approach the females. These males are often politely referred to as "sneaky copulators" (Zamudio and Sinervo 2000). An example of sneaky copulating is found in the horned dung beetle *(Onthophagus acumunatus)*. In this species, females are sequestered in burrows and males guard the entrance against rivals. However, when smaller and larger males are competing for a single female, smaller males will bypass the larger guards by digging secret tunnels to gain access to the female (Emlen 1997).

These examples are meant to convey two primary aspects of the male perspective on reproduction. First, the males of most species do not invest very much time or energy in offspring, and the continuous production of inexpensive gametes (sperm) gives males the potential of inseminating many females and fathering numerous progeny with little investment of energy. There are exceptions in species whose males produce spermatophores (large packages of sperm), which require a large energetic investment, but these are uncommon. Second, selection for access to females is strong, but there is a trade-off between putting effort into increasing access to females and putting effort into activities that enhance survivorship. The trade-off applies not only to frogs, dung beetles, and deer but also to males of our own species.

Allocating Energy

During the life span of an organism, its physiology must be geared to trigger several major life events at the optimal time, including a time to become reproductively mature, a time to reproduce, and ultimately, a time to die. To schedule and perform these life events, an organism needs energy. For mammals, energy comes from food resources that must be harvested, stored, and allocated among often competing physiological needs.

Most people have some favorite sweet treat. For some it is ice cream. For me, donuts and freshly brewed coffee definitely get my attention. The fate of the calories from my donut reflects a process that is fundamental to life history theory. The process involves the concept of allocation. Organisms are not homogeneous furnaces that burn energy equally from head to tail. Each component of the body has its requirement for optimal function. What would happen if I lived in an environment in which calorie-rich foods were not available whenever I wanted them? Not every component will always get what is optimally necessary. Energetic resources are often in short supply, and every type of tissue must find a way to muddle through with what is allocated.

As with time, energy that is used for one purpose is unavailable for other purposes. If I burn a log in my fireplace, I cannot use the same log to warm my neighbor's house. When I ingest calories, whether from a donut, a piece of fruit, or any other form of food, my physiology must decide the fate of the energy made available by the metabolic process. The same applies to stored energy such as fat that is broken down and used for energy. Fortunately, evolution has provided organisms with physiological mechanisms that make these allocation decisions in very efficient ways.

A calorie can be metabolized immediately or stored for later use, usually in the form of fat. What will become of the numerous calories provided by my donut? Many will surely be sequestered into fat cells, but what of the rest? Some will be excreted in waste, but the remainder will be left to perform many of the functions that keep me alive to crave sweet fried dough and caffeine. Should I fuel my immune system or some other aspect of my metabolism? If I were thirteen again, my body would also have to decide whether some of those calories should go to skeletal growth.

Different tissues and organs require different amounts of energy to function. For example, the brain accounts for about 20 percent of your basal metabolic rate (BMR), the amount of energy it takes just to keep you alive. Skeletal muscle tissue accounts for another 20–25 percent of your BMR. However, you have much more skeletal muscle mass than brain mass: per unit of mass, the brain is much more demanding than muscle. Other organs such as the heart, liver, and kidneys require energy, as do the immune and reproductive systems. But organs or systems cannot always have as large a piece of the energetic pie as they would like. As a result, most tissues exhibit what seem to be adaptive responses that are characterized by changes in their energetic demands. Such changes can be effected by adjusting the body's overall metabolic rate, adjusting the size of specific tissues, or adjusting the metabolic rate of selected tissues.

A Time to Die

The possibility of seeing another tomorrow is never assured. Indeed, in some organisms it is downright unlikely. Time, in contrast to food, cannot be harvested or stored. The amount of time available for an organism to grow and reproduce is partly a function of environmental hazards. However, even in the absence of external dangers such as predators or pathogens, organisms die on different schedules that seem to be somewhat preordained. For mice it is a few years, for humans several decades. The length of time organisms have in which to grow and reproduce, and the rate at which they senesce, seem to be incorporated into the physiology of their species (Rose 1991). It is almost as if different species have evolved specific times for their warranties to run out. But why senesce at all, and can senescence be altered? Several theories are currently under debate. Robert Kirkwood suggested what has come to be known as the "disposable soma" theory, proposing that organisms will invest in maintenance according to their life history prospects (Kirkwood 1999). That is, species with low extrinsic mortality (from predation, disease, and so on) should invest more heavily in somatic maintenance than species with higher extrinsic mortality.

A refinement of this idea involves the role of reproductive effort and associated costs to survivorship. Increasing investment of energy in reproduction seems to result in a faster rate of senescence. So why doesn't an organism simply play it cool and reproduce slowly? The answer seems to lie in environmental hazards. If members of a species with a high risk of predation, say rabbits, were in less of a hurry to propagate, they might not live to do so. For them, it might be a good idea to reproduce at high levels today just in case the fox came around tomorrow. However, this increased investment in reproduction seems to carry an inherent cost associated with senescence (Bell and Koufopanou 1986).

Life spans are finite; therefore, principles of allocation come into play. As mentioned earlier, the immediate evolutionary aspect of time constraints involves time allocation by individuals in everyday life. How much time shall I use to gather food versus searching for a mate? How much time should I invest in my older offspring if I am nursing or caring for a younger child? These decisions are crucial, and in humans they are especially crucial in populations that obtain their food from foraging or from agriculture based on manual labor. For males, the central issue of time allocation seems to be the trade-off between time spent augmenting survivorship and time used in reproductive effort. We'll discuss this in greater detail in later chapters.

How do we quantify life span? Is it the maximum recorded age of an individual within a species? How about the average age of death? Neither of these definitions is satisfactory. The first is subject to distortion effects of outliers: some atypical individuals in a population will live well beyond the expected age of death. The second doesn't provide a picture of how mortality varies across the lifetime. One way to get around this is to generate a mortality curve and determine the amount of time it takes for the mortality rate of a species to double.

Mortality curves are graphs of the chance of dying per year (y axis) over the years of life (x axis). Most mammals exhibit a U-shaped mortality curve with high annual mortality at the beginning and the end of life. The middle tends to be flatter. With long childhoods, extended adolescence, and a life span that exceeds our reproductive life (at least in women), it takes a relatively long time for the rate of human mortality to double, about nine years averaged across the life span. For mice, it is 0.3 years (Austad 1997a). Note that using the doubling rate allows us to compare species. As we near the end of life, the time it takes for our mortality rate to double shortens dramatically. Using the doubling time, one can observe a pattern that reflects the life span of that species.

Sex and Death

Up to now I have not mentioned sexual selection, and this has not been an oversight: sexual selection is part and parcel of male life history trade-offs. Probably no other aspect of evolutionary theory is so rooted in the basic principles of life history theory. As a graduate student, I was present at an argument that became quite heated. Was sexual selection a distinct form of selection, or was it the same as natural selection? Some argued one side, some the other. The dispute ultimately became a lawyerly exchange based on who had the best memory of *On the Origin of Species*. Both sides, I believed, were wrong, since the dichotomy they were assuming was false. While sexual selection differs from certain aspects of natural selection, it is still a form of natural selection.

As discussed in the previous chapter, nonrandom elimination of certain alleles in a population is accomplished by means of differential fertility and mortality. Natural selection is commonly viewed as increases in survivability as the result of some adaptive traits such as protective plumage or more efficient digestion—any characteristic that prolongs individuals' life spans and so increases their opportunities to produce offspring or to help their close genetic relatives do so. But what if a particular trait imposes significant increases in both fertility and mortality? If that trait, even while increasing mortality, results in greater lifetime reproductive success, it will be favored. Classically this is a central aspect of sexual selection. At their very essence, natural and sexual selection are forms of nonrandom elimination in which time and energy trade-offs influence survivorship, reproductive effort, and ultimately lifetime reproductive success.

Trade-offs between survivorship and reproductive effort are central to the evolution of human male physiology. Women also face trade-offs between survivorship and reproductive effort, but because of differences between male and female investment of energy in offspring, this particular trade-off is of greater relevance to the evolution of human males. In addition, understanding the life history trade-offs between reproductive effort and survivorship is crucial to gaining any insights into the relationship between specific male traits and female preferences for those traits. Nonrandom elimination in the form of sexual selection has been a central influence on the evolution of human male biology and life histories. If I could revisit the argument I witnessed in graduate school, I would propose that important components of sexual selection reflect nonrandom elimination (natural selection) that favors reproductive effort over survivorship.

Genes or Environment?

"All aspects of the phenotype are the product of interactions between genes and environment. The life history of an organism is one of the more comprehensive examples of a phenotype, for it results from processes stretching from molecular biology to ecology" (Stearns 1992). At this point it is necessary to introduce two important terms, genotype and phenotype. Genotype is the genetic complement that codes for a physical trait. Combined effects of genotypes contribute to complex physical traits. A phenotype is the physical manifestation of the combined contributions of genetic and environmental effects. Some traits are highly constrained by genetics and are under very little environmental influence. An example is the number of eyes in a mammal. The occasional mutation may arise, but no environmental change can result in any deviation from two eyes. Most phenotypes, though, are under significant environmental influence. Despite the public enthusiasm for finding "the gene" for various cancers or for behavioral tendencies such as aggression, even the most conservative genetic trait, the phenotypic expression of which seems to be isolated from environmental effects, is never expressed within an ecological vacuum.

Take, for example, phenylketonuria or PKU, a disorder that results from mutations in both alleles of the gene coding for an enzyme called phenylalanine hydroxylase (PAH). When synthesized in its proper form, PAH converts one common amino acid, phenylalanine, to another, tyrosine. Mutations in both copies of the gene for PAH result in an inefficient or inactive enzyme, causing concentrations of phenylalanine to accumulate with serious toxic effects, including mental retardation, organ damage, and ultimately death. At first glance, this would seem to be an example of a purely genetic trait that results in an unalterable destiny. However, even in this example of a person with mutations in both alleles of the gene, a change in the environment can drastically adjust the gene's expression, that is, the accumulation of phenylalanine. The change? The person merely has to stop consuming foods or beverages that contain phenylalanine, such as diet drinks.

J. M. Tanner's book *Fetus into Man* (1978) is a classic in human physiology and retains an honored place on my bookshelf. Within its pages is an extraordinary photograph of a set of male monozygotic twins, both sharing the same genotype. You'll note that I don't say "identical twins" even though that is the usual vernacular term for monozygotic twins. Despite these brothers' shared genetic makeup, their appearance demonstrates the influence of environment. One twin, raised in a nurturing household, appears robust and in good health, while his brother, separated from him at birth and a victim of severe neglect, is smaller and visibly less robust (Figure 2).

These dissimilar twins illustrate a frustrating aspect of the well-known argument about "nature versus nurture"—the relative roles of genes and environmental factors in shaping human characteristics—namely, that the correct answer is often "both." Certain physical traits, such as blood type, are completely the result of an individual's genetic complement, while green hair may be due to the purely environmental circumstance of walking under a falling can of paint. However, the overwhelming majority of physical characteristics, especially adaptations, result from interaction between genes and the environment. These adaptive traits are actually responses to the environment, and this responsiveness is known as "phenotypic plasticity." It is often advantageous for a phenotype to maintain a range of variation in the likely event that the environment changes. Droughts, food shortages, or other challenges often stimulate an organism's biology to alter the expression of a gene or a suite of genes. Hormones and other agents are the regulatory mechanisms that tweak the expression of genes in response to some environmental cue such as low blood sugar or decreases in fat.

The evolution of our species is characterized by our ability to con-

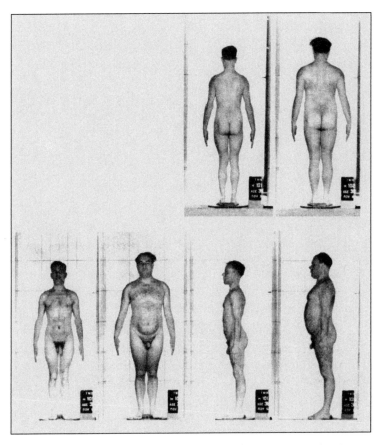

Figure 2 Monozygotic twins, one raised in a nurturing home, the other neglected.

form to radically different ecologies. We humans are quite good at adjusting our genotypes to produce an appropriate phenotype for the environment. Phenotypic plasticity, in combination with large brains, sociality, and a decrease in extrinsic mortality, has allowed our species to evolve into the remarkable primates we are.

3

The Ancestral Male

WITHIN THE HUMAN SPECIES, males, or least their gametes, are required for reproduction. But this is not true in all species. In fact, the first forms of life did not need sex at all. Bacteria reproduce asexually by simply doubling their genetic material and dividing into daughter cells. No muss, no fuss. Insects such as bees, and even some vertebrates, reproduce without male involvement. Whiptail lizards are all females, which reproduce parthenogenically. That is, mothers lay eggs that hatch and develop into daughters that are virtual clones of their mothers (Cole 2002). Clearly, the male sex has not been a necessity in the evolution of many organisms. So where did males come from, and why do some organisms need them? Sexual reproduction and the emergence of the first males occurred at least 2.5 billion years ago (Bernstein et al. 1981). Mammals came along as the age of the dinosaur was winding down about 65 million years ago. Among these furry, lactating creatures were the first primates, small squirrel-sized animals with unique skull features, forward-facing eye sockets, as well as larger brains (Sargis 2002). By the time primates came onto the evolutionary scene, sexual reproduction and the male sex were well established.

Sexual reproduction and the necessity of males puzzled Darwin. He wondered why, if evolution by natural selection was driven by the production of copies of oneself, an individual would dilute its heritable complement by mixing it with that of another individual. Darwin was not aware of genetics, but he knew that evolution required some mechanism of heritability. There are several theories regarding the evolution of sexual reproduction and the introduction of a male sex. Among these is the idea that the addition of a second sex provides increased genetic diversity to cope with an ever-changing environment. Asexual species, without this source of diversity, have a more limited range of genetic variation upon which selection can act, and are at greater risk of succumbing to modest changes in the environment and becoming extinct.

Spermatogenesis may be an important source of genetic variation as the result of errors in the replication of genetic material, one of the benefits of continuous generation of sex cells. The biologist J. B. S. Haldane argued that the production of sperm generates more mutations than the production of ova. He predicted that in primates, mutations in sperm should be about five times more frequent than mutations in ova (Haldane 1947). Recent evidence, however, has suggested a much lower rate of mutation during spermatogenesis than Haldane estimated. Hacho Bohossian and colleagues, who studied chimpanzees, gorillas, and humans, found very little difference between the mutation rates of male and female gametes. They suggested that perhaps mechanisms for repairing mutations during spermatogenesis are more efficient than previously thought (Bohossian et al. 2000). Kateryna Makova and Wen-Hsiung Li, who estimated human mutation rates using comparative data from chimpanzees and gorillas as well as more distantly related primates such as gibbons and siamangs, found a significantly higher alpha score, the number that reflects the male-female ratio in mutation rates, than that obtained by Bohossian and colleagues. The higher proportion of male mutations in humans reported by Makova and Li is in accordance with other studies and supports the central role of male DNA mutations as a significant source of genetic variation during human evolution. This controversy is surely not over, and future research will provide more evidence about whether males played a particularly important role in the evolution of the hominid lineage or if mutations in female gametes were equally significant. What is fairly certain is that about 6 million years ago a group of African apes emerged that were to be of central importance to the origin of modern humans, and that they placed us on an evolutionary path that diverged from that of our present closest relative, the chimpanzee.

Sexual Dimorphism in Hominids

Several million years ago in eastern Africa, within the present borders of Ethiopia, Chad, and Kenya, strange apes roamed the landscape. These creatures had small brains, by human standards, but they walked on two legs, more or less. Although there are large gaps in our knowledge about these ancient apes, analysis of fossil remains of some of them— *Ardipithecus ramidus, Australopithecus afarensis, Sahelanthropus tchadensis,* and *Kenyanthropus platyops*—places them very near the split in the hominoid lineage that would give rise to humans and chimpanzees. *Australopithecus afarensis* has presented paleoanthropologists with a remarkable picture of early hominid life. Despite its relatively small brain, not much bigger than that of a modern-day chimpanzee, and skull characteristics strikingly similar to those of chimps, *afarensis* was a card-carrying biped, although the curvature of its fingers and toes suggests that these individuals still spent a considerable amount of time in the trees.

One prominent feature exhibited by *afarensis* was sexual dimorphism in size: the males were about twice as large as the females. Modern *Homo sapiens* also exhibits some degree of sexual dimorphism in size, and there has been some discussion regarding possibly strong sexual dimorphism in other hominids. But the scale exhibited by *afarensis* is unmatched except in the modern orangutan and gorilla. *Afarensis* canine teeth were also very sexually dimorphic, with males having much larger canines than females. Moreover, *afarensis* males' canines were smaller than those of chimpanzees but much larger than those of humans. In mammals, canine teeth are common targets for sexual selection, that is, sex-specific changes that confer some reproductive advantage. Males under intense selection to compete for females usually exhibit large canines as well as large body mass compared with females.

Afarensis males were probably highly competitive with one another, most likely over access to females. How can we know this about an animal that has been extinct for more than 6 million years? The answer is analogy. If there is one strong and constantly reaffirmed relationship between physiology and social structure, it is that sexually dimorphic species are characterized by intense male/male competition, polygyny, and high male reproductive variance (Plavcan 2000). That is, only a few males get to mate with most of the fertile females, and those successful males will father most of the offspring. Given the undeniable sexual dimorphism in *afarensis,* both in size and in canine teeth, these other characteristics were almost certainly present as well. Other australopithecine species, such as *africanus,* also exhibit pronounced sexual dimorphism (Lockwood 1999).

What does all this mean in regard to the behavior of our male hominid ancestors? There is reason to be cautious in making such judgments. As J. Michael Plavcan and Carel van Schaik point out, sample sizes in the fossil record are usually very small, so that it is often difficult if not impossible to determine whether physical traits represent sexual dimorphism or extremes sampled from a normal range of variation (Plavcan and van Schaik 1997a). It can even be difficult to tell whether different body sizes represent different species or males and females within the same species. This has been an issue in studies of later hominids, specifically *Homo habilis* (Miller 1991; Miller 2000). With *afarensis,* the debate is still alive and well. Proponents of the single-species hypothesis point out that the level of dimorphism observed in *afarensis* does not exceed what is observed in primate species today. This may be so, but it is nonetheless extreme compared with other present and past hominoids.

If *afarensis* was extremely sexually dimorphic, are there possible explanations other than male/male competition? Evolutionary biologists have suggested that bigger, stronger male bodies might have evolved in response to pressure from predators. This hypothesis assumes that males are on the front line when dealing with marauding predators, but this is not always so. In addition, while body weight is often seen as a deterrent to predators, as in elephants and rhinoceroses, the defensive function of sexually dimorphic canine teeth is less apparent—although some primates, such as baboons, have canine teeth larger than those of leopards and often use them as effective weapons. Plavcan and van Schaik also note that only two of the four australopithecine species exhibit significant sexual dimorphism in body size. If we assume that all australopithecines were under similar risk of predation, this discrepancy appears to be evidence that predation pressure is not the source of larger male body size.

It is also possible that males have not been the target of selection at all. Perhaps a selective force has favored smaller female body size. It has been proposed that sexual dimorphism in chimpanzees might have decreased as a result of food resources being more scattered and female maturation being later than in other primates. The significance of delayed maturation is that female mammalian growth often ceases when reproductive maturation begins. Scattered resources and later female maturation, according to the anthropologist Steve Leigh (1995), might promote greater female/female competition. He proposed that if you reverse this logic when considering *afarensis*, females exploiting more evenly distributed resources, resulting in earlier maturation and less female/female competition, might favor a decrease in female body size, thereby increasing sexual dimorphism. However, this explanation, based on modest sexual dimorphism in extant primates, appears extreme when applied to the degree of sexual dimorphism seen in *afarensis*. In addition, the associations between the distribution of resources, foraging strategies, and degrees of sexual dimorphism implied by this idea are not found in observations of extant primates.

Sexual dimorphism is much less pronounced in *Homo sapiens* than in *afarensis* or in modern-day apes. Perhaps the emergence of modern humans, in conjunction with more sophisticated tools and language, shifted the realm of male/male competition from constant physical battle to the provision of food or other resources. In contemporary hunter-gatherer groups, as we'll see in a later chapter, men's hunting skill is at its greatest well past their age of peak physical strength and prowess. Perhaps increases in brain size emphasized and coevolved with the benefits of learned foraging skill in our ancient human male ancestors.

The First Humans?

At cookouts and parties, when people hear that I'm an anthropologist, they inevitably ask me two questions. What have I dug up lately, and when did the first "human" appear? The answer to the first question is easy. Nothing, unless you count the rocks in my lawn. As a biological anthropologist studying human evolution, I work with hormones. I don't do shovels. The second question is a bit more difficult. Whether or not *afarensis* and other australopithecines were human requires a definition of the term "human." Since this is not a book on philosophy, I will stick with measurable physical and life history traits. Using this set of criteria, I would say no, australopithecines were not human. I would postulate that one would have to look at our own genus, *Homo,* for any likely candidates. Members of the species *Homo erectus* were the first hominids to leave Africa and colonize parts of Asia. Their fossils have been found in China as well as Indonesia. Why did they have wanderlust? No one knows for sure. However, *Homo erectus* must have had several characteristics that enabled its members to trek such long distances and withstand changes in ecology as they went along. In general, *erectus* needed to be quite adaptable to the environment and able to tolerate random mortality from sources like predation and climate. Making lots of babies wouldn't have hurt their chances either.

In 1984 Kamoya Kimeau, a member of the research team of the anthropologist Alan Walker, made one of the most spectacular finds in anthropological paleontology. Kimeau discovered, along the western shores of Lake Turkana in Kenya, a remarkably complete *Homo erectus* skeleton, which the scientists named Nariokotome boy (Walker and Shipman 1996). This individual, probably male, was about five feet three inches tall and perhaps thirteen years old at the time of death, although some estimates have him as young as eight (Dean et al. 2001; Moggi-Cecchi 2001). Anthropologists have spent years examining Nariokotome boy and the stratigraphic surroundings in which he was found (Brown et al. 1985). For our purposes, two aspects of Nariokotome boy are most interesting: his size and his implied rate of growth.

Nariokotome boy was significantly larger than his australopithecine ancestors. Indeed, although he was short compared with modern humans in western industrialized societies, he was well within the range of human height variation in the modern world as a whole. Estimates of his age were drawn from several lines of evidence. First, the spaces between the growth plates of his long bones were virtually closed, suggesting that his long bones had stopped growing and he had reached his full adult size. Evidence of age was also obtained from tooth eruption patterns and dental development. To have attained his full height by the age of thirteen, Nariokotome boy must have grown faster than modern humans. To have reached full height at any age younger than thirteen implies an impressively fast rate of growth.

His brain was dramatically larger than those of australopithecines. Given the high cost of brain metabolism, *erectus* must have been able to exploit a rich resource base that allowed the maintenance of a larger, more expensive brain, larger, more expensive body size, and fast growth. Behavioral developments may have contributed to these morphological and life history changes. For example, the first solid evidence of the use of fire comes from *erectus* sites in Choukoutien, China, implying that *erectus* not only actively hunted for game, for which there is ample evidence, but had the ability to cook meat, thereby increasing the efficiency with which this species could exploit and utilize calories.

The role of males in this apparent increased exploitative power to obtain and utilize food resources is speculative. Were males supplying food to females to supplement their diets? We don't know. However, the metabolic requirements of *erectus* females were quite high compared with those of present-day primates and other hominid ancestors (Aiello and Key 2002; Leonard and Robertson 1997). It is plausible that females had become more efficient at obtaining food for themselves, or that they were receiving supplemental food from members of their social group. For *Homo erectus* to leave Africa and colonize new and often inhospitable areas in Asia, members of social groups probably played a role in carrying or caring for offspring. Mammalian ovarian function is very sensitive to energetic fluctuations and will decrease in response to high expenditure of energy, low caloric intake, or both. The success of *erectus* as colonizers surely involved an ability to adapt to changing climates as well as the development of behavioral strategies and physiological mechanisms that allowed females to maintain a reasonable level of fertility while dealing with uncertain environments and a mobile lifestyle.

Besides sharing food with females, how else could males have been involved? Perhaps by carrying infants or other loads, relieving the burden on females from time to time. This is speculative but based on some threads of logic. Less sexual dimorphism in *erectus* than in australopithecines implies less male/male competition and perhaps more male investment in females and offspring. Was *erectus* monogamous? Again, we don't know. It may be that the route of male reproductive effort had taken a significant turn from battling other males for access to females to helping maintaining a female's health, contributing to her fertility, and perhaps investing time and energy in the survival of offspring.

Real People

About 200,000 years ago, hominids that were virtually indistinguishable from you or me appeared on the African continent (Cann et al. 1987; Ruvolo et al. 1993). One of the most important developments in the evolution of modern humans was the ability to settle on every continent with the exception of Antarctica. This required life history attributes that allowed them to utilize time and energy efficiently in varied environments. Human physiology had to adjust to differences in the availability and quality of food, while maintaining and indeed increasing the rate of reproduction. Assuming that the physiology of the first modern *Homo sapiens* was virtually identical to our own, we can hypothesize that the modern humans leaving Africa 200,000 years ago had large brains, extended life spans, a long period of juvenile dependency, and reproductive maturation that did not occur until sometime in their second decade of life. Moreover, human lifetime reproductive output would have been much greater than is usual for mammals of our size. Taken together, these characteristics would have allowed humans to adapt to varied environments, tapping into rich sources of calories such as meat to provide the fuel for greater fertility. But what of the delayed sexual maturity? According to current life history theory, delays in maturity in organisms with determinate growth, such as humans, require decreased mortality caused by extrinsic factors like predation. That is, in order for maturation to be delayed, circumstances would have to allow humans to live a relatively long time.

Comparisons of mortality patterns between captive and wild chimpanzees and a human hunter-gatherer population illustrate the differences that probably gave humans the time to delay reproductive maturation, grow larger, and perhaps use their large brains to acquire and retain useful information about their environment (see Figure 3). In the wild, chimpanzees have several predators, including leopards, baboons, and even other chimpanzees (Boesch 1991; Watts et al. 2002; Wrangham 1999). Partly in consequence, they have higher mortality both as juveniles and as adults than members of a human foraging population such as the Ache.

Notice that even captive chimpanzees, though not in danger from predators, have higher mortality than humans at almost all ages. Chim-

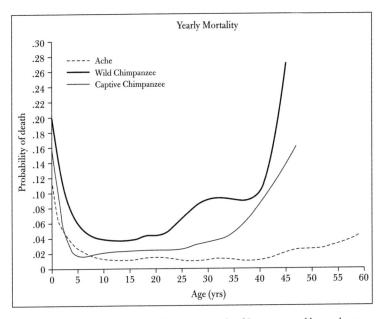

Figure 3 Mortality rates of wild chimpanzees, captive chimpanzees, and human hunter-gatherers (the Ache of Paraguay).

panzee evolution has selected for faster growth, earlier reproductive maturation, and more rapid senescence. Why is there a relationship between earlier or faster reproduction and increased rates of senescence? Biologists have suggested that earlier or more intense reproduction tends to divert energy away from physiological processes that may augment survivorship. Therefore, organisms that reproduce earlier or faster in response to greater risks of extrinsic mortality often grow old and die sooner than organisms that do not face similar risks (Austad 1993, 1997b). More on this later.

In order to evolve slower senescence, humans must have found a way to decrease extrinsic mortality. During our recent evolutionary past, predation was still a real danger. Even today, among hunter-gatherer groups, predation occurs often enough to warrant vigilance (Hill and Hurtado 1996). The most likely causes of the decline in extrinsic mortality involved cooperative defense. Observations of wild chimpanzees have suggested that sophisticated tools are not necessary to ward off predators such as leopards. In many organisms, having extra pairs of eyes, ears, and nostrils is a common way of defending against predation. One may speculate that tool use, weaponry, and larger brains may also have something to do with this, but that is placing the cart before the horse. For larger brains and sophisticated tools to evolve, the organisms had to grow at a slower rate and invest more energy in brains. To do so, the organisms needed time. We're back to senescence. Whatever the reason, humans have been able to fend off predators, and thus have been able to invest more time in growth and in expensive tissues such as large brains.

More time is available, but what about energy? Recall that those early humans also had higher fertility than other mammals of similar size—a change that would have required additional inputs of energy. Those inputs may well have come from increased food provision by males, most likely in the form of high-quality, difficult-to-acquire foods such as meat. Comparative data from several foraging groups illustrate the contribution of men to the group's diet. Beginning around the age of fifteen, young men begin to surpass young women in the caloric value of the food they supply, with their peak food contribution occurring between the ages of 30 and 45. The additional daily calories provided by males are a tremendous boost for energetically demanding processes such as reproduction in women. Whether this degree of provisioning by men was common during our evolutionary past is an open question.

But certainly, supplying additional food to women would have promoted fertility.

Why couldn't women simply hunt and acquire enough food for themselves? For one thing, they are often the primary caregiver for children. It is difficult to stalk and chase prey when you have an infant on your hip. Also, because of the high energy demands of reproduction, expending large amounts of energy in the pursuit of food would probably result in lower fertility. Peter Ellison and his research group have shown that even modest declines in weight or increases in energy expenditure can decrease women's levels of reproductive steroid hormones and decrease the chances of conception (Ellison 2001; Lipson and Ellison 1996). In women in western industrialized societies, light to moderate exercise or modest weight loss can compromise ovarian function (Lager and Ellison 1987, 1990). However, for most women outside these societies, agricultural labor and the everyday chores of life are sufficient to temper the production of hormones by the ovaries (Ellison et al. 1989b; Jasienska and Ellison 1998; Panter-Brick et al. 1993). So if early human women had done their own hunting and provided all their own food, their fertility would probably have suffered. As we have seen, in human groups that still live by foraging, the men consistently provide a significant number of calories, perhaps allowing females to maintain high fertility relative to human body size.

Hillard Kaplan and colleagues have proposed that the ramifications of male provisioning during human evolution may be at the core of many of the life history traits that make humans unique. For example, the human brain makes up approximately 20 percent of our resting metabolic rate; that is, it demands about a fifth of the minimum number of calories required to keep us alive with no activity. The brain does not alter its metabolic demand except under the most severe conditions of starvation. The evolution of the larger human brain was probably made possible by greater access to high-quality food resources. The ability to learn sophisticated hunting techniques and develop cooperative strategies for catching prey created further selection pressure to evolve a larger brain. Together, the benefits of hunting and the subsequent ability to become better, more efficient hunters may have been a key turning point in human evolution. Without males' provision of food, human evolution would have probably taken a different path (Kaplan et al. 2000).

Other implications of male provisioning include the evolution of human mating systems and strategies. Polygyny is the most common mating system among mammals and the predominant arrangement in human societies (Daly and Wilson 1983). However, contemporary human societies exhibit a broad variation in mating systems, including monogamy, serial monogamy, and even polyandry, as well as polygyny. In addition, the evolution of offspring that require substantial parental care over a long period of time has coincided with the development of increased male investment in mates and offspring, in the form of either provision of food or protection from predators or other males. Why did modern humans develop a system that seems so different from those of australopithecines and contemporary chimpanzees? A decrease in overt, physical male/male competition, less sexual dimorphism, as well as the evolution of distinct female reproductive strategies (Betzig et al. 1988; Hrdy 1981), surely contributed to making us the species we are today.

PART II

Human Male Life History

4

Stacking the Deck

Babies are expensive, economically, metabolically (for women), and in terms of care time. In species whose offspring require large amounts of parental resources and care, the parents' investment is quite high to raise youngsters to maturity. Females also have a finite amount of time to produce offspring. Turning out progeny that have little or no chance of reaching maturity would obviously not be a worthwhile strategy. Mechanisms that optimize the Darwinian return on investment in offspring in the form of greater numbers of grandchildren would have a significant evolutionary advantage. But as a parent, how does one actually optimize the return? One way in sexually reproducing organisms is to take some control over the number of offspring produced of a given sex. Why would this be advantageous? Let us consider an extreme hypothetical example. Say a population of organisms with equal numbers of males and females suffered a dramatic decline in the number of males because of disease, predation, or some other calamity. In consequence, males would be at a distinct advantage over females in the effort to pass their genetic material to future generations, since there would be very little male/male competition and males would be in great demand by females looking for mates. Thus females who produced male progeny would also have a clear evolutionary advantage. Without an active mechanism to produce males, a female would have to rely on luck. A mutant female able to control the sex of her offspring could conceivably produce a preponderance of males and therefore enjoy an evolutionary gain over other members of her cohort.

Does this sort of thing happen in nature? Yes, it does. For example, among wild baboons of Tanzania, male infants are more likely to survive to one year of age when there is little rainfall early in the year; female infants are more likely to live through that first year if they are among the first offspring produced by their mothers. Researchers found that social and ecological conditions favoring improved survivorship for a given sex were positively associated with the production of that sex. For example, mother baboons' early births were biased toward females, and when rainfall in January was low, subsequent births were biased toward males (Wasser and Norton 1993). Can it happen in humans? Maybe. After a major earthquake in Kobe, Japan, in 1995, the proportion of male births declined significantly from an expected value of 0.516 to 0.501 (Fukuda et al. 1998). Misao Fukuda and colleagues suggest that the widespread stress caused by the earthquake may have altered sperm motility and perhaps frequency of sexual intercourse, contributing to this ratio shift (Fukuda et al. 1998). It is unclear, however, how producing fewer male babies in such circumstances would be beneficial or adaptive.

What is certain is that being male involves higher mortality than being female. A small but consistent difference in mortality between males and females results in human sex ratios that are not quite what one would expect. A casual observer of human and animal communities might assume that they have about equal numbers of males and females.

In a world of sexual reproduction, this would seem to make perfect sense. The mathematician and theoretical biologist R. A. Fisher formalized this concept and examined why sex ratios in most organisms tend to be quite close to 1:1. Fisher and later scientists convincingly argued that arrangements that deviated from this ratio were evolutionarily unstable and would soon return to the state of equilibrium that includes roughly equal numbers of males and females. Fisher reasoned that parents should produce the same numbers of male and female offspring if the fitness costs and benefits were equal for each sex. If the sex ratio of young at independence deviated from 1:1, the parents who had produced more of the rarer sex would have higher expected lifetime reproductive success (Fisher 1930). However, Fisher's calculations were based on several assumptions that are rare in real-world situations. One Fisherian assumption was that natural selection was neutral, an unlikely scenario. It is possible, and indeed quite likely, that if selection favored one sex over the other, biased sex ratios would emerge. And in fact in many species the process of making males (or females) is seldom left to chance.

In numerous species, physiological mechanisms have evolved that manipulate sex ratios in response to environmental stimuli. For example, sex determination in some reptiles is influenced by the temperature of the sand in which the eggs incubate. In the American alligator, warmer sand increases the probability of producing a male, while in some turtles, the opposite occurs (Mrosovsky and Yntema 1980). Regulation of the sex of offspring in these reptiles is thought to be a response to population sex ratios. When there are few males, mothers will alter the temperature of egg incubation, perhaps by varying the depth at which they bury their eggs in the sand. Some female reptiles also sway their offspring's sex assignment through self-regulation of their own body temperature. *Eulamprus tympanum,* a species of Australian skink whose females give birth to their young rather than laying eggs, exhibits such a mechanism. Gestation at higher temperatures increases the probability of producing a male. It is likely that such shifts occur in response to social conditions: in the laboratory, females produced all sons when they were in the presence of all adult females (Robert and Thompson 2001).

What is the cue that triggers sex manipulation? Is it olfactory, visual, or some neurological calculation that involves many subtle environmental indications? The answer remains elusive. Members of some hermaphroditic species such as the clown fish, *Amphiprion percula,* can actually change their sex in response to the intensity of male/male competition. When male/male competition is high, males tend to change into females. When male/male competition is low, females tend to change into males (Shapiro 1992). Not many organisms go to such an extreme—but manipulating the sex ratio of offspring is not a matter of the occasional anomalous species, but a physiological trait that may emerge under significant selection pressure.

Another intriguing mechanism of sex-ratio adjustment involves stress hormones. Oliver Love and his collaborators hypothesized that female starlings should produce more female offspring in response to experimental elevation of a stress hormone (corticosterone). Elevating this hormone would mimic a well-established response to environmental, social, or energetic stress. Metabolically, female offspring are less energetically demanding to the mother. Therefore under conditions that elevate maternal stress hormones, more female offspring should be produced. Indeed, this is what was observed. Mothers with experimentally elevated corticosterone produced more female offspring because of higher mortality of male embryos. When males were born under these conditions, they tended to be smaller and less able to ward off infection (Love et al. 2005).

Skewed Ratios

The sex ratio of people between the ages of fifteen and sixty-four in the United States appears to be 1:1. However, this is not true when one looks at specific age classes or when examining the data with keen scrutiny. Newborns exhibit a slightly but consistently skewed male-biased ratio of 1.05:1, or 1.05 boys born for every girl. Americans older than sixty-four exhibit a female-biased ratio of 0.97:1, primarily because of the more rapid senescence and higher mortality of human males, which will be discussed in a later chapter. Not only in the United States but in many countries over the past century, the percentage of male births has consistently exceeded 51 percent (Parazzini et al. 1998). Male babies are more numerous even though more male than female fetuses are miscarried, especially in the early weeks of development (Kellokumpu-Lehtinen and Pelliniemi 1984). Boys are also more likely than girls to be born prematurely and less likely to survive their first year (Ingemarsson 2003). Thus in order to maintain equilibrium, it would be necessary to produce slightly more males.

Among adults, sex ratios in the United States and other countries are also slightly but consistently outside the expected range of variation around the 1:1 ideal (Allan et al. 1997; Moller 1996; Parazzini et al. 1998; van der Pal-de Bruin et al. 1997). Statistically, one would not expect an absolutely equal number of males and females because of random fluctuations. However, one can calculate the expected range of drift around the mean ratio, and the amount of variation observed is consistently outside that range. Shifts away from randomness are not static and can change. For example, for reasons that remain elusive, the percentage of male births, while remaining over 51 percent, declined notably between 1940 and 1995 (Nicolich et al. 2000).

Do these small changes in the relative number of male newborns indicate some selective force acting on human sex ratios? The jury is still out, but this has not discouraged researchers from pursuing the question from a theoretical perspective. The evolutionary biologists Robert Trivers and Dan Willard (1973) hypothesized that in polygynous organisms such as mammals, females should produce more sons and invest disproportionately in sons over daughters when resource predictability is high and maternal condition is good. This hypothesis has been difficult to test, and results have been mixed (Chacon-Puignau and Jaffe 1996; Gaulin and Robbins 1991; Koziel and Ulijaszek 2001). However, intriguing evidence of sex-ratio adjustment in humans in response to a disproportionate number of males or females has emerged. When the number of females declined in a preindustrial Finnish population, significantly more boys were born (Lummaa et al. 1998). This is not definitive evidence for sex-ratio adjustment in humans, but it would be odd for such mechanisms to be common in other mammals and vertebrates but absent in humans.

The anthropologists Ruth Mace and colleagues (2003) pondered whether variation in sex ratios might be related to overall fertility and mortality rates. They hypothesized that more girls should be produced in populations with high fertility and accompanying greater risk of maternal mortality. Their analysis of data from numerous countries supported this notion. In addition, Mhairi Gibson and Ruth Mace (2003) reported that rural Ethiopian women who produced more sons also tended to exhibit more muscle mass than other women, an indication of greater strength and better condition. Gibson and Mace suggest that since boys impose a greater metabolic cost on mothers (a topic to be covered later), only those whose bodies can "afford" having boys will tend to do so. The overall importance of these investigations is their indication that sex ratios in present-day society are sensitive to selective pressures and that interactions between biology and culture merit serious consideration.

How does the sex ratio get skewed? There are several possible ways. More ova might be fertilized by sperm containing the male Y chromosome, either because men produced more Y-carrying sperm or because certain aspects of the female reproductive system favored conception using Y-carrying sperm. However, changes in the number of sperm carrying a specific sex chromosome have not been observed in human males in association with unexpected sex ratios. Men who had fathered a disproportionate number of girls or boys revealed no evidence of an unusual number of X- or Y-carrying sperm (Irving et al. 1999). A comparison of the percentage of Y-carrying sperm in a sample of men and sex ratios at birth in several European countries showed a small but consistent discrepancy, with 50.3 percent of sperm being Y-bearing and 51.3 percent of births being male (Graffelman et al. 1999). However, since the balance of X- and Y-bearing sperm was within range of statistical expectation, the authors suggested that perhaps uterine conditions influence the sex ratios observed at birth. Indeed, since females bear most of the metabolic costs of reproduction and can absolutely determine the relatedness of offspring, it is likely that adjustment of sex ratios is influenced by female factors.

Sex ratios can be tweaked after fertilization, most likely through the disproportionate termination of pregnancies of one sex. More male than female stillbirths have been found to occur in some European countries (Byrne and Warburton 1987; Hassold et al. 1983). Caution is merited, though, since the manner in which nations record perinatal mortality can differ in significant ways that may affect comparative analyses (Graafmans et al. 2001). Nonetheless, it is fairly well established that male babies have lower survivorship than girls after birth (Ingemarsson 2003; Naeye et al. 1971; Stevenson et al. 2000).

Selecting Sons

Sex selection is not a rare occurrence in some societies. Cultural norms that promote the desirability of male offspring seem to be the driving force behind practices that favor the survivorship of sons over daughters. Sex selection through abortion of female fetuses and through infanticide is not uncommon and has been shown to have a significant effect on childhood sex ratios. It is very difficult to gather quantitative data about infanticide because of the obvious ethical and legal implications. However, an investigation in Hong Kong found that women who had already given birth to two daughters were significantly more likely to have sons. The third babies of women whose first two children were boys had a sex ratio of 0.94, those whose older siblings were a boy and a girl, 0.92, while those with two older sisters had a male-biased sex ratio of 1.37 (Wong and Lo 2001). This evidence is circumstantial and may possibly reflect factors such as selection for Y-carrying sperm, but Wong and Lo note that women who had already had two sons were no more likely to have a third than women whose earlier children were a son and a daughter.

In India and China, selective abortion of female fetuses has become a controversial issue. Selective abortion occurs with higher frequency in India and China than in other regions of the world (Bhat and Zavier 2003). Clinics that specialize in identifying fetal sex using amniocente-

sis and conducting selective abortions have proliferated in northern India. Indeed, in India the prohibition of amniocentesis has been suggested as a way to curb abortion for sex selection (Gorman-Stapleton
1990).

Disproportionate female infant mortality due to neglect of girl babies
or the diversion of resources to boys is also not infrequent. In rural
Peru, for example, Anne Larme (1997) found that girls were more likely
than boys to suffer from illnesses such as gastrointestinal and respiratory infections, but that despite their poor health, girls were less likely
than boys to receive medical attention that required payment. Similar
patterns have been noted in regions of India (Ganatra and Hirve 1994).

The long-term implications of cultural practices that skew sex ratios
are uncharted. However, it is virtually certain that having significantly
fewer females in subsequent generations is not a good idea for social
stability. In contemporary mainland China and India, two of the most
populous nations in the world, the long-term ramifications of increases
in the number of men who will never be able to marry or have relationships with women because of declining numbers of the latter are only
beginning to be realized. Potential future problems such as war, social unrest, and possible government consideration of culling the male
population through increased military recruitment or high-risk public
works projects are disturbing notions (Hudson and Boer 2004). The
implications of sex selection in favor of males are sure to generate discussion among public health officials, bioethicists, and others for years
to come.

5

Womb to Grow

B Y   S O M E   S T A N D A R D S ,   H U M A N   F E T A L   growth is not very different from that of other mammals or vertebrates in general. We've all seen the high school biology pictures in which early fetal pigs, chickens, and humans are almost indistinguishable. Human gestation is also not extraordinarily lengthy compared with that of other mammals of similar size, though for the mother it still requires a substantial investment of time and energy. Fetal growth is not linear, nor is it equal between various tissues. An inordinate amount of energy is devoted to fat and brain matter, and growth accelerates rapidly during the first two trimesters and slows dramatically as birth approaches. Compared with other species, humans dedicate a long time to childhood growth. Humans average about twelve years of slow, steady growth before they start reproductive maturation. This is an exceptionally long period even for primates, which grow more slowly than other vertebrates (Charnov and Berrigan 1993). The reasons for humans' slow steady growth are not completely clear, but it is probably rooted in lower extrinsic mortality as a result of greater parental care (Kaplan et al. 2000).

Development before birth and during childhood is dominated by the needs of maintenance and growth, an indication of the importance of survival to young males compared with later years, when investment in survival is often tempered by the needs of reproduction. Devotion of time or energy to reproduction will not begin until adolescence. Although long, slow childhood growth is indicative of fewer environmental risks than are faced by other primates, the initial investment of time and energy solely in survival is understandable because, even during gestation, there are significant risks.

Fetal growth is not without pitfalls since mother and fetus are unique entities with individual life history agendas. Estimates of early pregnancy loss range from 30 percent (Wang et al. 2003) to 60 percent (Wood 1994) to 75 percent (Wilcox et al. 1988). The risk of spontaneous abortion is especially high during the first trimester and may involve the elimination of conceptions that contain genetic abnormalities. Mechanisms of detection of such abnormalities by the mother's body are poorly understood, but the life history agendas of the fetus and the mother exhibit a tension that persists throughout infancy and childhood. For example, maternal/fetal differences over the allocation of glucose may be evident in various conditions such as gestational diabetes. These risks are not unique to male offspring, but they do reflect the beginning of a lifelong balancing act involving the management of available resources (Haig 1993).

In addition to gestation, infancy and childhood are life stages that are riskier than later periods in life. In all human populations, from the poorest shanty towns to the most affluent urban areas, infants are at substantial risk compared with older children. Mortality is universally higher between birth and the fifth birthday than later in childhood. Why? Perhaps, given young children's small size, basic energetic demands result in insurmountable challenges to their immune systems.

Infants and young children are exceptionally susceptible to infectious disease, most commonly gastrointestinal illnesses that cause diarrhea and the resulting dehydration, the most common source of child mortality in the world. The extended period of growth in infancy and childhood can be viewed as an attempt to address these dangers by devoting time and energy to growth and maintenance. Indeed, growth and maintenance are often intertwined: some aspects of growth are crucial for developing a robust physiology with an immune system that can cope with environmental challenges.

Human males exhibit higher mortality than females throughout their lives. At conception, male zygotes appear to be produced in greater numbers than female zygotes, but to be of poorer quality. An investigation of induced abortions in Finland showed that sex ratios of fetuses were highly skewed toward males during the first few weeks of gestation but then steadily became less skewed in later weeks (Kellokumpu-Lehtinen and Pelliniemi 1984). Whatever mechanism was culling males did so early on, perhaps in response to genetic abnormalities. Among low-birthweight and prematurely born infants, too, males are more likely to die than girls (Ingemarsson 2003; Stevenson et al. 2000). More research is needed to determine what is killing off disproportionate numbers of male fetuses and male newborns; what is certain is that high mortality is a male characteristic, not only at this stage but also during adolescence and in old age. It seems that being male is not good for your health.

The Mutant Female

A male's existence begins when a Y-bearing sperm fertilizes an X-bearing ovum. Afterward, many hormonal signals that guide development can be traced to the activation of genes on the Y chromosome. Male physical differentiation from females involves obvious morphological changes, such as genital development, but also others that are much more subtle, such as neurological growth and organization. The line between males and females is somewhat murky at the earliest stages since the sexual organs of both sexes emerge from the same initial structures.

Physiologically, the development of male fetuses can be likened to the creation of mutant females. This is a simplification of a very complex process, but it emphasizes the subtlety of the differences that separate males from females. All human fetuses are sexually undifferentiated at the beginning. And whether or not they carry the Y chromosome, unless they are subjected to certain specific hormonal influences at specific times, they will develop into individuals that *appear* to be female. For example, individuals with a genetic disorder known as androgen insensitivity syndrome have the male XY complement of chromosomes but develop external female genitalia. This rare condition is the outcome of a genetic anomaly typified by dysfunctional or absent hormone receptors that are meant to sense the presence of male sex hormones such as testosterone. Without the ability to detect these hormones, which are responsible for changing an undifferentiated fetus into a male phenotype, the fetus will develop apparent female genitalia (Ahmed et al. 2000). Indeed, this disorder is usually not diagnosed until the teen years when the supposed girl, lacking ovaries and a uterus, fails to menstruate. In order to make a male, the default trajectory for fetal development must be altered.

To trace the development of a male fetus, let us start in the middle, with the testes. Colorfully named Y chromosome genes such as "desert hedgehog" direct the production of testes determining factor, which coaxes the growth of testicular tissue, specifically sperm-producing Sertoli cells, around the eighth week of development (Bitgood et al.

1996; Koopman et al. 1991; Pelliniemi et al. 1993). Leydig cells begin to develop around the sixth week, producing testosterone and dihydro-testosterone (DHT), two potent androgenic steroids that initiate and nurture penile development and the fusion of the labial folds to form the scrotum. Around the eighth week another agent, Müllerian inhibit-ing substance (MIS), initiates the degeneration of the Müllerian ducts, structures that in females would become fallopian tubes and other bits of internal reproductive organs. With the Müllerian ducts out of the picture, another set of structures, the Wolffian ducts, are permitted to form the male internal reproductive tracts.

It is very difficult to gather quantitative data on fetal hormones be-cause of the obvious constraints of having your research subject inside another person. However, by carefully drawing blood from the umbili-cal cord, Paolo Beck-Peccoz and colleagues (1991) surveyed 114 normal fetuses during different periods of development. They noted that sub-units (specific fragments) of the pituitary hormones LH and FSH were significantly lower in males than in females between week 17 and week 24 of gestation. It is unclear why this difference occurs, but the finding underscores the sex-specific nature of hormonal function during fetal development.

In the last weeks of gestation, LH and FSH levels decline in both sexes, with testosterone rising in males and staying elevated until imme-diately before birth, when it drops to the same levels found in females. The purpose of the rise and fall in testosterone in male fetuses is uncer-tain. Perhaps it involves sex-specific changes in brain organization.

DHT, which is produced by the Leydig cells along with testosterone, is a form of super testosterone. Although the amount produced is only a fraction of the amount of testosterone, DHT has a much greater ability to bind and activate testosterone receptor cells. One particularly important cluster of androgen receptors located in the genital buds responds to DHT, stimulating the clitoral bud to develop into the penis. The importance of DHT is evident in a rare condition found in a small population in the Dominican Republic, presumably due to the proliferation of a genetic mutation. The condition is known locally as "penis at twelve" and clinically as five alpha-reductase deficiency. Fetuses with this mutation have the male combination of X and Y chromosomes but do not produce five alpha reductase, the enzyme necessary to convert testosterone to DHT. Without DHT, fetal penile development doesn't occur, and the infant is born with a female genital appearance despite having an XY genotype and functioning but undescended testes. At birth these children appear to be female, but when they reach puberty their testes begin producing large amounts of testosterone. The testosterone stimulates the growth of the clitoris into a penis and thereby changes the supposed girl into a boy (Imperato-McGinley et al. 1979). This condition is very rare, but it illustrates the importance of androgens and other hormones in organizing external maleness.

The Developing Brain

Despite behaviors that may indicate otherwise, the development of human male and female brains is basically identical. However, there are subtle contrasts in function and organization that can trace their origins to the womb. Many of these differences are manifested in contrasting cognitive abilities as determined by a multitude of psychological tests. It must be cautioned, though, that most of the evidence we have is from animals and that the limited information we have about humans is gleaned from westernized populations. This is not to suggest that this information should be ignored but to encourage new directions of research.

Testosterone binds to numerous receptors within the brain. Evidence of interaction between testosterone, the brain, and behavior is substantial in rodents, birds, and nonhuman primates. We know far less about humans, primarily because of the ethical and logistical difficulties of conducting research on human fetuses. Nonetheless, various case histories and comparative studies can shed some light on the role of sexual differentiation in the male brain.

The question of what happens when testosterone interacts with the brain during development has been a controversial topic, with its implication that behavioral differences between boys and girls or between men and women are to some extent influenced by biological factors before birth. Differences in behavior between boys and girls have often been attributed to cultural or social influences that mold sex roles. Certainly social factors have a strong influence, but more detailed analyses have shown that societal influences are but one aspect underlying sex-based differences. It would be naive to assume that millions of years of mammalian evolution involving differential selection pressures on males and females would result in a single brain type that is not selected to deal with sex-specific challenges. However, determining the selection pressures involved in male brain evolution and identifying the targets of selection are not easy tasks. It would not be unreasonable to suggest that the evolution of the male brain has been subject to selection pressures that address the optimal trade-off between neurological aspects related to survivorship and those related to reproductive effort. Would this involve selection for a tendency to exhibit sex-specific behaviors? It probably would.

A full discussion of the morphology and function of the human brain is not the purpose of this book. But there are key landmarks and structures that deserve special attention, either because they reflect some unique aspect of male physiology or because they regulate crucial features of growth, maintenance, or reproduction. I will restrict the discussion to differences in the size of the two brain hemispheres, the primary connection between the hemispheres, and the hypothalamus, a core area of neuroendocrine control.

LEFTIST TENDENCIES

The left hemisphere of the brain is often smaller than the right in male fetuses, while in female fetuses the brain tends to be more symmetrical (de Lacoste et al. 1991; Hering-Hanit et al. 2001). In the male brain the planum temporale, an area associated with verbal and language processing, is larger in the left hemisphere than in the right—and boys and men tend to do less well on verbal tasks than girls and women (Deacon 1997). Girls and women show greater symmetry in the planum temporale (Kulynych et al. 1994). Other brain structures such as the corpus callosum, the bundle of fibers connecting the left and right hemispheres, appear to be less robust in males. In addition, some evidence from studies of primates suggests that there may be more testosterone receptors in the left cortical hemisphere than in the right, a difference that may be related to cortical asymmetry (Sholl and K. R. 1990). So why are male brains so asymmetrical? Do males have an inferior left hemisphere, or does the right grow disproportionately?

Sex steroid hormones almost surely play a role in sex differences in brain structure. We know that testosterone has a "masculinizing" effect on the brain. However, the role of testosterone is indirect: it is converted to estradiol, which is what actually changes cell structure. But estradiol is also present in female fetuses, so why doesn't it masculinize their brains as well? The answer lies in a substance known as alpha-fetoprotein (AFP). AFP is found in both male and female fetuses but is virtually undetectable in adults. Its function is somewhat unclear but it has been proposed to play an important role in sexual differentiation of the brain. AFP binds to maternal estradiol within the developing female brain and keeps it from entering the neural cells, thereby protecting the female brain from the masculinizing effects of estradiol (Bakker et al. 2006).

Why would males evolve a smaller left hemisphere? This is very difficult to understand empirically. Perhaps males represent the ancestral state and it is females who have evolved symmetry. That is unlikely, though, because testosterone is now believed to have evolved later than estrogens (Thornton 2001), so the effects of testosterone on brain morphology would have emerged after the evolution of the female morphological pattern. Adaptive explanations for male brain asymmetry are problematical to fathom. Is it somehow advantageous to be verbally challenged or less capable of recovering from brain injuries or strokes? Unlikely. Perhaps there is some metabolic or behavioral aspect that provides an advantage. If so, evidence has not been forthcoming. What can be stated with some credibility is that asymmetry has its costs. It is probable that the asymmetry of the male brain is not adaptive at all but an unavoidable epiphenomenon resulting from the production of testosterone. If this is indeed true, then what are the advantages of testosterone? We will consider this question soon.

Bridging the Gap

The largest bundle of neural connecting fibers that allow signals to pass between the left and right cerebral hemispheres is the corpus callosum.

The corpus callosum stretches sagittally (front to back) from the frontal cortex to the posterior sections, running almost the entire length of the brain. In cross section, the corpus callosum takes on the outlined shape of a mushroom cap, with bends at both the anterior (front) and posterior (rear) ends. The anterior bend is known as the genu while the posterior one is called the splenium. Anatomical examinations of fetuses (26–41 weeks into gestation) revealed significant differences between the sexes in the size, area, and shape of the splenium. Females were noted to have a wider and more bulbous splenium than males. In addition, females exhibited greater total corpus callosum area (Holloway and de Lacoste 1986). Replication of these tests has not yielded consistent results, especially in reference to corpus callosum area and size (Hwang et al. 2004). However, sexual dimorphism in splenium shape has been supported, with females indeed exhibiting a more bulbous splenium as adults (Clarke et al. 1989) and a thicker overall corpus callosum (Achiron et al. 2001). Why the inconsistent findings? Most likely the disparities are the result of measurement methods. Some researchers relied on gross anatomical examinations while others used magnetic resonance imaging (MRI) and ultrasound. Sample sizes are also an issue, as are difficulties in identifying the boundaries of the corpus callosum. A more recent method that fully utilizes the potential of MRI to examine brain tissue in three dimensions supported the finding that females have a larger posterior section of the corpus callosum than males (Davatzikos et al. 1996). Does this mean that females are more capable of shuffling neural information back and forth between the two hemispheres? Perhaps.

Comparisons of the isthmus region of the corpus callosum, which includes the fibers that connect the temporal (side) and parietal (top) regions of the brain, reveal more definitive differences between males and females, with women exhibiting greater isthmus tissue area (Steinmetz et al. 1992). More subtle but telling differences are evident in the composition of fibers within the isthmus. Associations with sex differences in speech and language abilities might be a distinct result. Females consistently perform better on verbal tasks and recover more easily from strokes and lesions that affect cortical regions that are important to speech, such as Broca's area (Aboitz et al. 1992; Diamond 1991; Wisniewski 1998).

Again, the selection pressures, if any, that led to sex differences in corpus callosum morphology are not understood. One could speculate that selection pressure for verbal ability and the resulting ability to communicate socially was greater in females than males. But social communication is arguably as important to boys and men as to girls and women. More likely, less hemispheric brain matter results in less fiber investment to connect the two halves. What would be the advantage of such an arrangement in males? None immediately comes to mind. The most likely explanation involves trade-offs with other male hormone effects that may indeed be advantageous.

GRAND CENTRAL STATION

Walking into Grand Central Station in New York City, one cannot ignore the vast number of people scuttling back and forth, often in what appears to be a chaotic fashion. But of course, it is not chaos. Travelers have destinations and schedules are met. From the outside Grand Central Station can be easy to miss—and yet it obviously plays a major role in the organization of the city. The hypothalamus, the center of neuroendocrine function and buried in the deepest recesses of the brain, is the body's Grand Central Station. Despite its minute size, the hypothalamus is a focal point for the control of reproductive, metabolic, and behavioral endocrinology. It is not really a structure, but a compression of neural cells that operate in a unique fashion. Indeed, its name simply describes the location of the neural cells (under the thalamus). Many nerve fibers stretch across the brain and from other parts of the body and terminate at the hypothalamus.

The hypothalamus translates neural messages into hormonal communication. How? By having specialized neural cells secrete a hormone called gonadotropin-releasing hormone or GnRH. GnRH is released in a pulsatile fashion, mirroring the on/off nature of the firing of neurons, into a small conduit leading to the pituitary gland. This short-lived hormone, lasting only several minutes, stimulates the pituitary to secrete LH and FSH in both males and females. The hypothalamus also secretes analogous hormones that target adrenal and thyroid functions. Later we will revisit the hypothalamus and GnRH in exploring a condition that allows a glimpse into their role in developing and organizing male life histories.

The hypothalamus has been studied intensely by those interested in sex differences in the brain. In animals such as rats, ferrets, and monkeys, the hypothalamus exhibits clear sexual dimorphism in both structure and sensitivity to testosterone exposure. In rats, exposure to testosterone *in utero* is an absolute necessity for the development of male-specific adult sexual behavior (mounting and thrusting). Until recently, evidence for sexual dimorphism within the human hypothalamus has been scant. Initial investigations reported significant adult sexual dimorphism, specifically within the preoptic area of the hypothalamus, with males exhibiting about 2.5 times greater neuronal mass and 2.2 times as many neurons in this area than females (Swaab and Fliers 1985). Men also exhibit significantly more androgen receptors in the hypothalamus than women (Fernandez-Guasti et al. 2000).

It is not surprising that the initial investigations reported sexual dimorphism in the human hypothalamus. In some birds, the hypothalamus actually changes its structure during adulthood in response to testosterone and the change affects the birds' singing and mate attraction (Balthazart 1991). While similar morphological changes have not been observed in adult men in response to testosterone, it is quite likely that the human male hypothalamus undergoes changes during fetal development in response to exposure to hormones. The story of the hypothalamus needs many more chapters. Comparisons of size and morphology are only the first steps toward identifying the potent effects of this powerful organizing area of the brain.

THIS IS YOUR BRAIN ON TESTOSTERONE

Testosterone is considered virtually synonymous with being male. But how much truth is in its reputation? Does testosterone actually alter male brain matter to create a different organ from the female brain? Direct observations of the effects of sex steroids on human fetal brain development are rare for obvious ethical reasons. Evidence from nonhuman organisms does support the suspicion that testosterone has an organizational and activational role in fetal brain growth. "Organizational" refers to actual changes in neural morphology that underlie behavior, while "activational" alludes to stimulation of preexisting structures.

One strategy of investigation is to measure the hormone levels of fetuses and see what transpires during their development. Gina Grimshaw and colleagues measured prenatal testosterone in girls and boys via amniotic fluid and conducted follow-up studies on the children's development ten years later. They found that higher testosterone levels were associated with right-handedness and with greater reliance on one brain hemisphere in verbal test performance (Grimshaw et al. 1995). Not exactly compelling evidence but intriguing in that the fetal hormonal environment does seem to have biological and perhaps behavioral effects down the road.

Other investigations have focused on unintentional exposure to hormones, observing the effects of natural and clinically induced conditions in which fetal brains were inadvertently exposed to androgens. A well-documented example involved the administration of diethylstilbestrol or DES. During the 1960s DES was administered to pregnant women who were at high risk of miscarriage. However, DES also had an unintended androgenic effect on developing brains (Hines and Gorski 1985). That is, DES seemed to mimic testosterone. Several studies using a battery of psychological tests were conducted to determine whether DES had indeed had some masculinizing effects on the brain. Sons born to mothers who had received at least one month of DES treatment during pregnancy, tested from preadolescence to adulthood, exhibited more right-handedness and more pronounced reliance on one brain hemisphere (Reinisch and Sanders 1992).

In a related study, women who had been exposed to DES for at least five months *in utero* exhibited greater asymmetry of brain function during verbal processing tests than control women. Their performance on the verbal tests suggested that exposure to DES had affected their female brains in a manner similar to what is found naturally in male brains (Hines and Shipley 1984). The likelihood that DES causes actual changes in the structure of the corpus callosum or the cortex structure is small, although it is entirely possible that DES alters the efficiency with which these structures operate, perhaps in the ability of neurons to fire. These findings also intimate that prenatal exposure to DES has effects on brain activation even in adulthood.

A concern about the DES studies is that these synthetic hormones may not be sufficiently similar to testosterone and other androgens. This is a somewhat minor concern since it is clear that DES activates testosterone receptors. But nonetheless, these compounds are not endogenous, that is, they are not produced by the body itself.

A fascinating characteristic of steroid hormones is the similarity of their molecular structure. Relatively small changes in their natural synthesis caused by errors in enzymatic function can have profound effects. Take, for example, the production of glucocorticoids, steroid hormones produced by the adrenal glands in both males and females and involved in energy mobilization and responses to stress. A genetic condition known as congenital adrenal hyperplasia (CAH) results in the production of androgens instead of glucocorticoids in female fetuses. These mismanufactured hormones cause genital masculinization, including enlargement of the clitoris and partial or total fusion of the labia. One may recall the effects of testosterone and DHT on normal male genital development. This condition is not diagnosed until after birth; once diagnosed, it is easily treated by removing all or part of the defective adrenal gland or rendering it functionless and administering lifelong cortisol treatment. An investigation of whether CAH has any masculinizing effects on the brain during fetal development found no significant differences between twenty-two women with CAH and age-matched controls in vowel listening tasks and other tests. A survey of several other investigations did not reveal any significant differences in cognitive function between CAH and control girls (Berenbaum 2001).

Clearly, the role of androgens in masculinizing brains is far more complex than simple exposure of receptors to these hormones. But why did DES seem to have a masculinizing effect and not CAH? The answer may involve variation in the psychological and verbal tests being administered, subtle differences in the biochemistry of these steroids, or the extremely labile nature of behavioral development. It probably has something to do with all these factors.

It is clear from animal studies that the brain does contain receptors that respond to testosterone and other sex steroids. Primate studies involving the administration of labeled testosterone and estradiol have shown sex differences in brain sensitivity to these hormones both within the cerebral cortex (the outer neural portion of the brain) and in subcortical regions such as the hypothalamus (areas involved in emotion and spontaneous behavior) (Bonsall and Michael 1992; Bonsall et al. 1990). Moreover, it has been shown through genetic engineering of mouse genes that genetic coding on the Y chromosome can affect brain organization independent of exposure to testosterone (Carruth et al. 2002). Changes to brain structure are evident in the increase in lateralization in response to testosterone in males. However, the expression of these hormonally based neural differences in everyday cognition in adults is much less clear. Perhaps the psychological tests now available simply do not test for differences in function that are being elicited by testosterone. It may be useful for future researchers to conduct longitudinal investigations of males who exhibit some form of genetically based hormonal deficiency as well as those with genetic conditions that keep them from producing testosterone (Huhtaniemi 2002a, 2002b; Martens et al. 2002).

WHY THE DIFFERENCES?

The evolutionary significance of neurological sexual dimorphism is poorly understood. This is not for lack of thought. Indeed, entire journals are devoted to the evolutionary significance of human behavior and the accompanying physiology. The difficulties lie in (1) developing testable hypotheses that examine a particular question with some reasonable degree of validity, (2) devising quantitative techniques to test those ideas, and (3) identifying valid neurological differences between the sexes that may have been subject to sex-specific selection pressures. The brain is an immensely complex organ that is challenging to examine and understand when the subject is communicative and available for research, as in the case of adults. Understanding the neurological nuances of fetuses will require much more sophisticated techniques than those available today. Nonetheless, it is important to generate valid hypotheses rooted in a thorough understanding of adaptation and evolutionary theory.

In addition, the range of variability in brain function and morphology is staggering. To tease out variation in association with sexual development, one must understand and control for the range of neurological plasticity of the human brain. Even when there are gross errors in neural organization caused by some endocrinological or other mishap, the brain seems to function relatively well. It also exhibits an ability to adapt to injury or illness. Indeed, one of the most striking demonstrations of brain plasticity occurs in patients with extreme epilepsy who are treated by hemispherectomy, in which parts of one brain hemisphere are removed. Remarkably, after a recovery period these patients are able to return to their normal activities. The sight of an MRI showing empty space where half the brain should be in a person who is functioning normally illustrates how little we actually know about the brain and how much we still have to learn.

Assuming that natural selection has influenced the evolution of the male brain in a sex-specific manner is not a broad leap of logic. The question is what those selection factors were and how they are expressed in the organization of the brain *in utero*. Selection for sex-specific patterns of brain organization and function is probably due to the importance of regulating sexual behavior and motivation as well as behaviors related to risk assessment, time management, and so on. As will be evident in later chapters, male reproductive effort is reflected by behavioral factors in many vertebrates, including humans. It should therefore not be surprising that fetal brain development involves sex-specific differences in behavior that influence reproductive effort even if they do not help survivorship. Recall from previous chapters that sexual selection involves selection that favors reproductive effort over survivorship.

But why make alterations in a brain *in utero* that may prove to be permanent? Why not select for more plasticity and allow for greater freedom to alter brain physiology in later years? Well, the brain does indeed maintain a significant amount of plasticity, especially during childhood and almost certainly extending into adolescence. But for the brain characteristics that do seem to be set during fetal development, why not make them more labile, more capable of changing? Why is it important to make these changes so early in the organism's lifetime? First, it is important to remember that, male or female, the brain is responsible for the most changeable of phenotypes, behavior. In male humans, unlike some birds and reptiles, testosterone does not trigger a suite of stereotypic aggressive or sexual behaviors. In humans, testosterone affects the propensity to engage in broad classes of behaviors that we associate with being masculine, a somewhat subjective description to say the least.

Perhaps in the case of testosterone and other sex steroids that are important to sex-specific brain organization, physiological constraints limit the range of neural plasticity. Hormone receptors often exhibit an initial stage of sensitivity that is short lived. The receptors can change their sensitivity to hormonal stimulation in response to initial exposure to that hormone. Once the hormone is withdrawn, subsequent exposure to it evokes an enhanced response. Perhaps the surges of testosterone in male fetuses prime neurons to become more responsive to testosterone exposure later in life, thereby making the individuals more prone to behaviors typical of adult men such as aggression and competitiveness.

The selection pressures leading to testosterone sensitivity may have been less strong for early humans than for their hominid ancestors such as *Australopithecus afarensis,* as reflected by the high degree of sexual dimorphism in that species. High mortality resulting from risky behaviors has been found among young men in a variety of populations from hunter-gatherers to urban dwellers, suggesting the likely universality of male brain function, behavior, and risk assessment. Such high rates of mortality would not persist unless the behaviors that imposed these mortality costs also conferred some fitness advantage. The fact that this period of high mortality coincides with peak lifetime testosterone levels is almost surely not coincidental. As more sophisticated neurological assessment techniques are developed, it is virtually certain that new and previously unsuspected associations between androgens and brain development will emerge.

An Early Start

Mammalian males, whatever their species, are often larger than females. This is true in humans. The size discrepancy appears early. The body mass of newborn boys is, on average, 2–4 percent greater than that of newborn girls. This difference extends across numerous populations. The anthropologists Richard Smith and Stephen Leigh, in one of the most extensive surveys of neonatal sexual dimorphism, found male newborns to be significantly larger than female newborns in a variety of cultural and environmental conditions (Smith and Leigh 1998). The difference is small compared with later sex differences in growth and development, such as those during puberty, but it is consistent.

The size differences between human male and female fetuses are somewhat generalized. Abdominal circumference, head width, and head circumference are all significantly larger in male fetuses at 20 and 30 weeks (Hindmarsh et al. 2002). Long bone growth is not significantly different between males and females. The mechanisms responsible for sexually dimorphic growth during fetal development are unclear, but several possibilities such as differences between males and females in the secretion of growth hormone and insulin-like factor have been suggested (Geary et al. 2003). However, investigations attempting to link GH with fetal growth have yielded conflicting results.

Body composition is also an issue of interest since greater body mass in males may be the result of adiposity, muscle mass, bone, or a combination of various tissues. It is extremely difficult to examine the effects of testosterone on muscle development and growth in human fetuses, but comparative data from rats and pigs suggest that testosterone is necessary to sensitize muscle receptors for future steroid-induced growth. In gerbils, when female fetuses gestate between two males, thereby increasing their exposure to testosterone, they exhibit 16 percent more production of motoneurons (neurons that connect to skeletal muscle) in muscle cells than female fetuses gestating between two females (Forger et al. 1996). Hormone levels in newborns have been shown to have a small but significant relationship with sexually dimorphic muscle strength during childhood. Interestingly, no relationship was found with testosterone, but girls with higher neonatal progesterone levels later had less muscle strength while boys with higher progesterone later exhibited greater strength (Jacklin et al. 1984).

Another issue that may be related to greater male fetal size is higher male infant mortality, and perhaps increased maternal mortality associated with bearing sons (Beise and Voland 2002; Helle et al. 2002). The greater energetic demands of a male fetus may make it more vulnerable to mortality or morbidity in the womb. Interestingly, a recent investigation found that women carrying male fetuses ingested 9.6 percent more calories, 8.0 percent more protein, 9.2 percent more carbohydrates, 10.9 percent more animal fats, and 14.9 percent more vegetable fats than women carrying female fetuses. The cue for this dietary behavior is unknown although the investigators speculate that the metabolic stimulating effects of fetal testosterone may be contributory (Tamimi et al. 2003).

When one thinks of a baby boy, spermatogenesis and raging hormones don't immediately come to mind. And yet, while their hormones aren't exactly raging, male infants are not totally quiescent from a hormonal standpoint. Immediately after birth there is a distinct but poorly understood rise in testosterone levels that dissipates after several months (Andersson et al. 1998). Moreover, between six and twelve weeks of age, boys exhibit pulsatile secretions of LH, and to some extent FSH. Also, inhibin B, a hormone secreted by Sertoli cells in the testes that aids in regulating FSH, is produced in significant quantities in newborns. But again, the production of inhibin B declines and is very low throughout much of childhood until the onset of puberty.

It is unclear why such a hormone surge exists in newborn boys. One suggestion is that exposure to testosterone after birth is important for complete differentiation and formation of the genitalia. In a small clinical study (three patients), infants without this rise in sex hormone levels were found to show signs of hypogonadism, that is, underdevelopment of the penis and scrotum, which was successfully treated with the administration of exogenous hormones (Main et al. 2000).

We know that the sexual and behavioral delineations of the human male begin before he leaves the womb, even if many of the mechanisms remain to be fully understood. It is likely that selection has acted to prepare males for the sex-specific developmental and environmental challenges that lie ahead. One of the important challenges to biologists who study human evolution is to distinguish between male traits inherited from our hominid ancestors and those reflecting significant selection pressures that exist today. As we move on to adolescent development, evidence will help to clarify which aspects of male life history are still subject to environmental selection pressures.

6

Getting a Life

AFTER BIRTH, THE BABY BOY'S behavior and physiology undergo a number of adjustments. His mother is no longer providing direct nutrients through the umbilical cord, so he must allocate time and energy among various physiological needs, mostly related to growth and maintenance. In hunter-gatherer populations, mothers and other family and community members play important roles in gathering the resources that young boys will need to survive childhood. Issues related to nursing, weaning, and even foraging for their own food and other resources are central challenges. Environmental hazards related to pathogens and accidents are significant sources of mortality and morbidity, not only among forager populations but in other contemporary societies. The first few weeks and months after birth are among the most perilous times, particularly for boys, whose infant mortality rates are higher than those of girls (Stevenson et al. 2000; Maconochie and Roman 1997; Zaldivar et al. 1991).

In infancy and childhood, with their hazards and challenges, male physiology devotes most of its resources toward survival. Reproductive function in childhood is virtually nonexistent. The hypothalamus is generally quiescent, and there is little or no gonadal activity. However, the lack of reproductive function is not evidence that life history trade-offs are nonexistent.

A Mother's Son

The behavioral and physiological changes that occur with the infant boy's introduction to the outside world involve a number of cooperative and negotiated strategies by mother and son. The newborn boy faces several primary challenges. First, he must grow. There is an initial spurt of growth during the first year of life followed by continuous, steady growth until the onset of puberty. In addition, he must gain the resources necessary to maintain this growth, and he must be protected from threats such as predators and pathogens.

One of the most basic questions is how much time is spent with the mother. Human infants and children require a tremendous amount of care and provisioning, and parents must cope with the trade-off between caring for their children and making an effort to produce more children. Are there sex differences between boys and girls regarding their relationship with their mother? To get a sense of conditions that may have been reflective of the majority of human evolution, we can revisit data from hunter-gatherer studies.

Observations of the amount of time spent with mothers among !Kung San foragers of Botswana revealed that infant boys (0–80 weeks old) consistently spent less time in direct contact with their mothers than girls did. When not in direct contact, they were with other caregivers or being monitored from a short distance (Konner 1976). Boys may spend fewer hours with their mothers because of greater male independence, mothers' preferential treatment of girls, or both. Among the !Kung, greater independence by male toddlers may be contributory. Similarly, among the Ache, boys tend to exhibit more independence than girls between the ages of two months and four years (Hill and Hurtado 1996), although no sex differences were noted in an assessment of language, motor skills, and personal/social development (Kaplan and Dove 1987). Cross-cultural studies, however, have largely failed to find consistent sex differences in the age at which toddlers begin to walk and exhibit independence (Stanitski et al. 2000; Yaqoob et al. 1993).

Breast feeding is the most direct method of transferring energy from mother to child. Its contribution to the infant's well-being includes not only direct caloric investment but also the transfer of various immune factors to the infant's immune system. Interestingly, recent evidence has shown that breast feeding is associated with significantly lower risk of upper respiratory infection in newborn girls but not boys (Sinha et al. 2003). The cause of this difference is not readily apparent, although one may speculate that if boys are more independent or nurse less vigorously, they may receive fewer immune factors. Indeed, among Hutterites, a community in which mothers regularly nurse their infants, daughters were weaned later than boys (Margulis et al. 1993). An investigation of Mexican-American infants, however, found girls exhibiting greater signs of infection in the form of wheezing in association with more breast feeding (Wright et al. 1989). Besides effects on immune function, some evidence suggests that breast feeding benefits boys' mental development more than girls', although the beneficial effects found were modest (Paine et al. 1999).

Environmental influences such as nutrition have a huge impact on growth patterns. Not surprisingly, boys (and girls) who suffer nutritional deficits during childhood are shorter and grow at slower rates.

For example, in a comparison of groups of children in Guatemala, Mayan boys of low socioeconomic status (SES) were consistently shorter than Ladino boys of higher SES, and the Mayans exhibited slower growth rates as they approached adolescence (Bogin et al. 1992). In addition, an examination of skeletal height of contemporary Guatemalan boys and girls and the excavated remains of ancestral populations showed that changes in height mapped onto periods of extreme economic, ecological, and political stress (Bogin and Keep 1999). This is not surprising, and little difference was found between boys and girls. However, significant differences are common in the manner in which boys and girls obtain and allocate energy.

A study of rural Nepalese populations found that boys did not work as hard as girls during childhood and adolescence. On average, girls worked 5.8 hours per day, twice as long as boys. The differences were primarily in the amount of domestic work (house maintenance and cooking) and economic work (selling items). Moreover, boys spent more time playing and studying than girls, although there were no significant differences in the time spent sleeping (Yamanaka and Ashworth 2002). There was also no significant difference in the number of calories ingested. This pattern is not uncommon and is primarily the result of the preferred status of boys in numerous societies and the sexual division of labor that often burdens girls with more work from an early age.

The Teen Years

In general, the onset of reproductive maturation signals the end of steady childhood growth in both boys and girls. When the reproductive hormones begin to be produced, energetic resources are diverted from skeletal growth to reproductive function. In boys, the predicted trade-off is between childhood growth and pubertal investment in reproductive effort.

To test for such a life history allocation in boys, a useful, albeit unethical, experiment would be to induce puberty early on, thereby forcing boys to begin allocating energy toward reproduction instead of growth. While this experiment is not possible, it does occur naturally in the form of a condition known as central precocious puberty (CPP), which occurs in both boys and girls and is characterized by the early awakening of hypothalamic reproductive function, well before adolescence (Kaplowitz 2004). This condition causes children as young as five years to undergo early pubertal development. It is often treated with hormones that block the effects of premature hypothalamic activity. During the teen years, treatment is ended and the person is allowed to undergo puberty.

According to life history theory's predictions about allocation of energy, untreated CPP boys should exhibit shorter stature because their energetic resources are being diverted to reproductive maturation. However, the evidence is mixed. CPP is associated with impaired growth and short stature, and attempts to reverse CPP do seem to ameliorate this effect. Hormone treatment is associated with some increase in height, but it is difficult to assess whether the final height is significantly different from the height patients would have reached without treatment (Couto-Silva et al. 2002; Lazar et al. 2001).

Adolescence is the period of transition from childhood to adulthood. The timing of this transition is an extremely important issue in life history theory, for it signals, in many organisms, the end of growth and the beginning of reproductive investment. Why is the timing of sexual maturation so important? What exactly changes physiologically and what are the mechanisms? After growing at a steady rate throughout childhood, why do adolescent boys undergo such a dramatic growth spurt? Why is boys' growth spurt later than that of girls? During adolescence, sexual dimorphism, the size difference between males and females, increases significantly. In addition, body composition diverges, with boys becoming more muscular and less fatty than girls. What selective forces shaped these growth patterns, and what are the implications for future health and life issues?

One cannot discuss puberty without talking about hormones. Indeed, endocrinology plays an essential role in life history biology. Although trade-offs and reaction norms are predicted under various circumstances, how does one go about testing predictions about the physiology underlying life history theory? It has been suggested many times that because hormones play a central role in translating genotype into phenotype, they are a valuable target of investigation (Stearns 1989; Ketterson and Nolan 1992). The endocrine system affects all aspects of life history, including the timing and function of growth, maintenance, and reproduction. The strategy of studying hormones to investigate life histories has been used for some time in other vertebrates but has only fairly recently to be utilized in humans (Finch and Rose 1995; Ellison 2003).

Male gonadal function is regulated by the reproductive neuroendocrine system, which includes the hypothalamus, the pituitary gland, and the testes (see Figure 4). From here on I will refer to this cast of characters as the HPT axis. Although the HPT axis is the central endocrine controller that regulates many aspects of reproductive function, other hormones are also involved, such as cortisol, insulin, leptin, and thyroid hormone, as well as a variety of other substances. Indeed, the line between what is considered part of the endocrine system and what is considered part of the neurological system can be somewhat arbitrary and

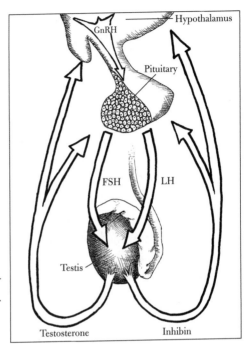

Figure 4 The hypothalamic-
pituitary-testicular axis.

confusing. For this reason many physiologists refer to the entire collection of neurological and endocrine structures and substances as the neuroendocrine system, paying homage to the manner in which they act in concert. What we'll discuss here is a streamlined version of male reproductive endocrinology and maturation.

Basic endocrinology is often compared to a television or radio broadcast. Signals are produced in a remote location and transmitted in all directions. These signals are detected and read by radio or television antennae. In this analogy, the broadcaster is the hormone-producing gland, the signal is the hormone, and the antennae are the target receptors. However, even with an antenna, one must be tuned to the proper frequency to detect a particular signal, such as a specific radio or television station. Hormone receptors, in general, are sensitive or tuned to a specific hormone.

Within the HPT axis there are several signals and broadcasters. The main broadcasters are the hypothalamus, the pituitary, and the testes. Gonadotropic releasing hormone, or GnRH, is secreted in a pulsatile fashion, one burst every few minutes, by the hypothalamus. GnRH pulses are usually first evident at night but gradually assume the normal pattern of daily pulsative activity.

When GnRH is produced, it travels down a series of microcapillaries through what is known as the hypophyseal portal. In essence, this portal is a circulatory passageway that leads directly to the pituitary gland. It's fortunate that the trip is short since GnRH begins to break down in the bloodstream within a few minutes. You may have noticed that the name of a hormone, in general, is an attempt to describe the hormone's primary action. Once GnRH reaches the pituitary, it lives up to its name. It stimulates the release of a class of hormones known as gonadotropins, namely luteinizing hormone (LH) and follicle-stimulating hormone (FSH).

Up to this point, I could have been describing either male or female reproductive endocrine function. From this point on, however, the sexes diverge somewhat.

Since GnRH reaches the pituitary in bursts, it is not surprising that the rest of the hormones downstream also exhibit a pattern of pulsatile secretion. LH and FSH, released into the bloodstream by the pituitary, make their way to target cells and receptors in the testes. LH stimulates the production of sex steroids, particularly testosterone and DHT, by specific receptors on Leydig cells. FSH is involved in the initiation and maintenance of sperm production by Sertoli cells.

Most hormones are quite similar in molecular structure. Only subtle differences separate them, yet their respective functions are quite well defined. This suggests that hormone function is very conservative from an evolutionary perspective, changing very little over time. This characteristic has been good news for diabetics who, before the widespread availability of human insulin, were treated with insulin from pigs. Most hormones have a deep evolutionary history, especially steroids. For this reason, among others, it is difficult to control the HPT axis in clinical research. Even if one blocks the production of a peptide or blocks a receptor, it is likely that another, related substance will remain unaffected and will stimulate the target cells. This is becoming less of a problem for researchers thanks to highly specific blockers and inhibitors.

The HPT axis operates much like a furnace thermostat in your house. As the temperature rises, the warm air triggers the thermostat to shut off the furnace at a predetermined temperature. As the temperature drops below that preset temperature, the thermostat restarts until the air again reaches the target temperature. This cycling process is what is called a negative feedback loop. To complete the negative feedback loop that is characteristic of the HPT axis, testosterone circulates back to the hypothalamus and inhibits the synthesis of GnRH. When testosterone levels drop to their minimum, GnRH is again produced. Other, less well understood hormones include inhibin, which is secreted by the testes and downregulates FSH production. This is the basic standard reproductive hormone package for the average adult male. However, reproductive maturation is characterized by a significant range of variation, both between individuals and between populations. What causes this variation, and what is its significance in the evolution of male life histories?

The Male Maker

After the initial burst of hormone activity in early infancy, not much happens from a reproductive hormone standpoint throughout childhood. We see low levels of GnRH, testosterone, and gonadotropin as a result of extreme hypothalamic sensitivity to the negative feedback effects of circulating steroids such as testosterone and estradiol. That is, the hypothalamus is basically on high alert for testosterone and its close relative, estradiol. As soon as either of these steroids is detected in the bloodstream, the hypothalamus slams on the brakes for GnRH production, and boyhood continues. However, at around the age of twelve or thirteen, the hypothalamus begins to be more tolerant of testosterone. Consequently, testosterone levels start to rise and the physical transition from boy to man begins.

Why does the hypothalamus lose its sensitivity to testosterone? This is not completely clear, but the adrenal gland may be involved. The adrenal gland, located just above the kidneys, is best known for its secretion of a class of steroid hormones known as glucocorticoids, of which cortisol is of primary importance. Glucocorticoids play vital roles in metabolism, energy utilization, and responses to stress. However, all steroids are very similar in molecular structure. Only minor enzymatic changes are necessary to produce a testosterone-like androgen instead of cortisol. In normal daily function, the adrenal glands produce small amounts of androgens such as androstendione. In its synthetic form, androstendione is marketed as Andro and used by some athletes as a bodybuilding supplement. During late boyhood there is a steady rise in adrenal androgens, probably due to general physical growth of the adrenal gland. This rise in adrenal androgens, often referred to as "adrenarche," may play an important role in initiating hypothalamic desensitization to testosterone.

Typically, testosterone ebbs and wanes throughout the day, with concentrations being significantly higher in the morning than in the evening, although the daily fluctuation may diminish with age (Bremner et al. 1983). Variation among individuals is considerable, with salivary testosterone levels in healthy men exhibiting tenfold differences (Schurmeyer and Nieschlag 1982). Investigations involving frequent blood sampling (approximately every ten to twenty minutes) have also uncovered a vast range of daily individual variation in testosterone and LH levels. In the course of a single day, some men exhibited seemingly random declines in testosterone to levels that would have qualified them for a diagnosis of hypogonadism. Some men exhibited typical daily fluctuations, while others did not. Unlike testosterone and LH, FSH was relatively invariant (Spratt et al. 1988).

Hormonal manipulation in rodents and nonhuman primates has revealed a significant range of variability between species as well as a detailed picture of how the components of the HPT axis interact with one another and with other endocrine systems. It is not possible to conduct similar experiments on human subjects, but various naturally occurring clinical disorders, although rare, provide insights into the details of male reproductive endocrine function (Jameson 1996; Layman et al. 1997).

A particularly useful model involves men with idiopathic hypogonadotropic hypogonadism (IHH). Men with this condition basically have a broken hypothalamus. Various mutations in the GnRH-transcribing gene within the hypothalamus result in complete or partial inability to produce GnRH and to activate the entire HPT axis. In essence, IHH males represent a GnRH knockout model that allows independent manipulation of hypothalamic function, GnRH availability, pituitary production of gonadotropins, and gonadal steroid levels (Crowley et al. 1991). Research on IHH men has confirmed many of the functions of the negative feedback system in the HPT axis. But several unexpected results indicate that the HPT system is sensitive to environmental conditions that may suggest the potential programming of male reproductive function.

In addition to testosterone's role in suppressing production of GnRH and LH, estradiol, a hormone more often associated with women but normally found at lower levels in men, also exerts significant influence on HPT function. For example, greater conversion of testosterone to estradiol within fat cells underlies low testosterone levels and hypogonadism in obese men (Zumoff et al. 2003). In addition, estradiol is suspected to contribute to the negative feedback effect on the hypothalamus (Hayes et al. 2000).

The significance of these results from an anthropological and life history perspective lies in the possible role of body composition in regulating male reproductive neuroendocrine function. Men whose energy resources are invested in greater adiposity (fat) exhibit greater conversion of testosterone to estradiol as well as hypothalamic and pituitary inhibition. As will be discussed later, changes in the allocation of investment in bodily tissue between adiposity and lean muscle tissue are often associated with shifts in the allocation of energy between survivorship and reproductive effort, with testosterone and estradiol playing potentially crucial regulatory roles (Bribiescas 1996, 2001a).

A second outcome of the IHH investigations casts light on what may be hormonal programming during adolescence. Briefly, programming involves changes in the sensitivity of hormone receptors in response to exposure to a particular hormone during a specific period. Hormonal programming is common and has been demonstrated in various endocrine systems and experimental models (Feigelson and Linkie 1987; Page et al. 2001). IHH patients are initially diagnosed when they fail to undergo pubertal changes such as the development of secondary sexual characteristics like pubic hair and penile growth. But not all patients are the same. Some IHH males exhibit no pubertal development into their second decade of life, while others manifest some pubertal development only to regress back to a prepubertal stage. Clinicians have dubbed these two conditions classic and adult-onset IHH. Some of the differences between the two groups suggest that hormonal programming occurs during adolescence. Men with adult-onset IHH, who were exposed to some of their own endogenous testosterone during adolescence, have been found to be more responsive than those with classic IHH to hormone treatment after the onset of the condition (see, e.g., Spratt and Crowley 1988).

Similar programming of reproductive hormone function has been suggested to occur in human females in response to improved environmental conditions and availability of food during adolescence (Ellison 1996). While the role of hormonal programming in male adolescents remains to be fully elucidated, variation in HPT function in association with ecologically driven differences in energy availability has been noted. For example, undernourished Kenyan male adolescents exhibit later pubertal LH and FSH increases than well-fed controls living in affluent neighborhoods of Nairobi (Kulin et al. 1984). Later pubertal maturation in rural than in urban boys in Zambia further indicates that nutritional status contributes to the timing of the onset of adolescence (Campbell et al. 2004).

Adolescence may be a key period in the life history of human males when an adult range of hormonal function is established, thereby potentially affecting reproductive strategies throughout later life. Factors that may contribute to hormonal variation and programming during adolescence and perhaps during fetal development include energy balance (Jean-Faucher et al. 1982; Rhind et al. 2001), disease (Spratt et al. 1993), and other constitutional stresses.

Why does this programming seem to occur in males? Perhaps it is an attempt to predict future environmental conditions and adjust hormonal function accordingly (Barker 1998; Kuzawa 2005). As we'll see in a later chapter, testosterone varies quite a bit between populations living under different energetic conditions. This variation may stem from conditions during adolescence that set receptor sensitivity for adulthood. Perhaps adjustments to the sensitivity of hormone receptors are attempts to optimize hormonal function in light of changes in the availability of energetic resources. But then why doesn't the body simply maintain plasticity throughout life? This is unclear, but it may be the result of some basic physiological constraint that remains to be defined.

Over the Hump

During adolescence boys undergo a distinct growth spurt that is indicative of reproductive maturation. If we were to view the growth patterns of males and females from birth to adulthood, we would see a general, steady increase in body size, with a small acceleration at the end, signifying the attainment on average, of a larger body size by males (Figure 5). But the *rate* of growth, that is, the amount of growth per unit of time, presents a more informative picture of the changes that are taking place when a male experiences reproductive maturation. Boys undergo

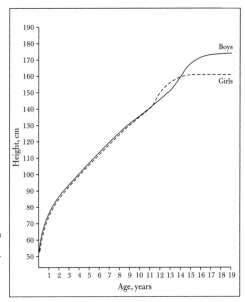

Figure 5 Cumulative growth in healthy boys and girls.

a growth spurt that is later and more pronounced than that of girls (Figure 6). Growth spurts are not uncommon among primates and tend to be greater in larger members of our order, such as ourselves (Leigh 1996); hence it is not surprising to find them in both male and female humans. The question of interest for us here is what contrasts in male and female life histories result in differences in the timing and amplitude of growth spurts.

In both males and females, the growth that occurs during adolescence is distinct from childhood growth. During the prepubescent years, bone growth is steady, with little or no input from sex steroids. However, adolescent growth spurts are a function of the hypothalamic awakening that results in the sudden increase in the production of sex steroids. Under the influence of testosterone in males and estrogen in

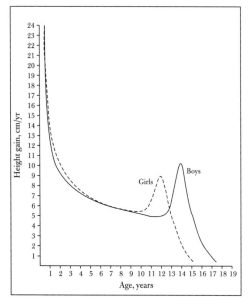

Figure 6 Rates of growth in healthy boys and girls.

females, this type of growth is sudden and involves growth that is sensitive to sex steroids, such as pelvic development in girls. As indicated in Figure 6, boys go through their growth spurts approximately two years later than girls, and once they start their period of accelerated growth, they grow faster than girls.

The accentuated growth spurt in adolescent boys contributes to the sexual dimorphism seen in adult humans. Whether this is the result of contemporary selection pressures or simply a relic of our evolutionary past is unknown. Sexual dimorphism is common in primates and mammals in general, so this may not be so unusual (Plavcan 2001). Indeed, as noted in an earlier chapter, sexual dimorphism appears to be associated with male/male competition (Plavcan and van Schaik 1997b). The differences in the timing of female and male growth spurts may be the result of accelerated female growth spurts, delayed male spurts, or both. One possibility is that male growth spurts are delayed because of the risks associated with exhibiting adult male physical characteristics.

If we turn our attention to orangutans, we can glimpse the types of selection pressures that might result in a sudden acceleration in growth and the potential benefits of delaying this growth. Male orangutans come in two flavors, large and small. Some males grow very large, to the point that their sexual dimorphism is the most extreme among the great apes. The large males also develop secondary sexual characteristics such as facial flaps and throat sacs. The smaller males do not develop these characteristics. The reasons for the evolution of these two morphs remain to be fully explained, but some aspects are clear. Once a male orangutan starts to exhibit signs of getting larger and acquiring secondary sexual characteristics, he becomes the target of serious aggression by other large males. The transition period is very risky, when the youngster isn't large enough to defend himself effectively, but is nonetheless on the large males' radar screen as a potential rival (Utami et al. 2002).

Perhaps in our own evolutionary past, male/male physical aggression was serious enough to exert a similar selection pressure. For a male, the transition from juvenile to adult involves a jump into the competitive arena. As discussed previously, mammalian male reproduction is primarily limited by the availability of females, either because females are scarce or because females prefer some males over others. Such circumstances inevitably result in males viewing other males of reproductive age as potential rivals. Among orangutans, some males have opted out of the competitive arena by remaining small. Perhaps they avoid direct confrontations with larger males. Being in transition between juvenile and large adult is a risky state, one that should be minimized for the sake of survival; and when this transition does occur, it should happen quickly. Hence the evolution of a delayed growth spurt relative to that of females, but one that is more pronounced when it occurs. Also, the delay may provide time for garnering energetic resources to make the eventual jump as rapid as possible.

A Life of Risk

Flipping through the multitude of cable channels, it is not difficult to come across programs dedicated to "extreme sports." Whether the particular sport involves jumping off a bridge while attached to an oversized rubber band, snowboarding down a Himalayan mountainside while guzzling a green soft drink, or attempting to ride a skateboard on a metal handrail above the granite steps of a local city hall, the predominant cast of characters is made up of boys and young men. When things go wrong in these pursuits, bones snap, knees bend at absurd angles, and it becomes quite clear that, despite their daring, boys are not immune to the effects of gravity and physics. Without modern emergency medicine, I doubt that many of these extreme sports enthusiasts would live to celebrate their eighteenth birthdays.

Teenage boys tend to do stupid things that cut their lives short. It doesn't matter if they are suburban teenagers in Toronto or hunter-gatherer boys in the forests of South America. Mortality due to accidents and aggression rises dramatically once a boy enters adolescence and does not start to subside until the end of his third decade of life. Hormonal changes are intricately involved in the phenomenon of higher mortality due to risk-taking behavior, although the exact physiology of the effects of hormones on risky behavior remains murky. Why would such behavioral patterns have evolved? What is the benefit of becoming less sensitive to danger? Why does good judgment take such a dramatic hit during this period? Part of the answer lies in the basic male trade-off, survivorship versus reproductive effort. However, before invoking evolutionary or life history theory, we have to identify the question that is pertinent to this discussion and address alternative factors that may contribute to risky behavior and the resulting increase in mortality.

It has been suggested that risky behavior by young men is simply a recent aspect of contemporary society. Violence on television, violent video games, and other media influences are seen as contributors to reckless behavior by male teens. While these influences are worthy of serious consideration and may contribute to some violent or risky activity, teenage girls, who are exposed to the same media influences, do not experience the same increase in mortality as their male peers. In addition, the prevalence of male risk-taking behavior in other societies that do not have easy access to television, films, or video games suggests that media influences are not the primary driver of young men's risky behavior and increased mortality.

The anthropologists Martin Daly and Margo Wilson examined mortality rates among male Canadians and found a sharp increase in mortality around the age of twenty that continued until just before the fourth decade of life. The primary cause of this increase was external factors such as accidents (Daly and Wilson 1983). There is some indication that mortality from risky behavior and accidents may have been accentuated during the last hundred years or so. According to a multi-population study covering the past hundred years, men in their early twenties have exhibited a dramatic increase in mortality attributed to accidents, perhaps because of the more lethal consequences of risky behavior involving more modern aspects of life such as motor vehicles and firearms (Kruger and Nesse 2002). But is this simply the case of modern urban youths run amok? Data from Ache hunter-gatherers suggest not. Mortality is also at its highest around the age of twenty among Ache males (Hill and Hurtado 1996).

But how would such a pattern of high mortality be maintained by natural selection? Remember that selection favors behaviors and physiological mechanisms that maximize lifetime reproductive success, so in attempting to explain the selection for risk taking, we would predict that, on average, males who engage in risky behaviors have greater lifetime reproductive success than those who don't. Does empirical evidence support this notion? For humans, it is unclear. However, male vertebrates consistently exhibit high mortality during their initial period of competing with other males for mates. Elephant seals, red deer, frogs, all show high male mortality in early adulthood. These same males have the potential to reap huge fitness benefits by engaging in risky behaviors and strategies.

Males' reproductive success exhibits much greater variation than that of females because of the physiological constraints associated with reproduction. Therefore the potential payoffs for males of braving certain dangers for the sake of reproduction are quite high despite the risks to survivorship. For example, Xavante men of Brazil exhibit reproductive success ranging from zero to almost two dozen offspring (Salzano et al. 1967). It is probable that the men with more children must hunt more to feed their families, and that the more one hunts, the more one encounters the dangers associated with hunting. So what's the benefit? Hillard Kaplan and Kim Hill have reported that for Ache men, success at hunting is largely a function of time investment. The more time spent

pursuing game, the greater return of meat (Hill et al. 1985). They also note that Ache women, who often have several sexual partners, tend to identify successful hunters as probable fathers of their children (Kaplan and Hill 1985).

Does any of this explain why selection would have favored a tendency for young men to take risks? Does lighting himself on fire or jumping off a building somehow make a man more attractive to women? I certainly hope not. However, there may be some male-bonding aspects at work. Being one of the group and accepting challenges may be a male characteristic that has come down from our ancestors to present-day societies, although the specific challenges have changed. Instead of jousting as knights once did or fighting with clubs as Ache men used to do, men in modern industrialized societies accept dares. Go figure. It may be that males in our evolutionary past who did not take the occasional risk did not fare well in gaining resources or the attention of females. Staying home under the covers, while surely comfortable and potentially good for survivorship, is not likely to be beneficial for reproductive fitness.

Why don't we see these risky behaviors in girls? To some extent we do. It is not uncommon to see female athletes pole-vaulting, skiing, or boxing. Women in foraging societies certainly expose themselves to hazards when they go out into the countryside to gather food and other resources. However, a key difference may be that girls and women generally do not seem to seek out risky tasks, especially those with no tangible benefit. And ultimately, female risky behaviors do not result in the same spike in mortality that we see in young men. Females have less to gain from a fitness perspective from taking risks to attract multiple mates. Since women bear much of the burden of childcare around the

world, they also have more to lose from engaging in dangerous behavior: bad outcomes from risky acts would surely have a detrimental effect on their dependent offspring.

Problem Child

Men and boys make up about 50 percent of the global human population, but in the United States and many other places they account for over 85 percent of the violent crime committed (Daly and Wilson 1983). This pattern begins at adolescence, suggesting that it is related to some aspect of puberty. Is it physiological? Is it purely sociological? How does the behavioral evolution of aggression and violence fit into the life history of a human male? Are there simple physiological aspects to violence that can be quantitatively measured and tested? Before addressing any of these issues, we need to realize that the concept of aggression is difficult to define and measure in a consistent manner. We may know aggression when we see it: it is easy to recognize, for example, in the face of a battered child or woman victimized by male violence. The difficulty lies in attempting to quantify aggression. One problem is that victims of male violence may be reluctant to come forward, fearing retribution or further harm. Therefore, it is not unreasonable to assume that the prevalence of male violence is often underestimated.

But what of the mechanistic causes? Hormonal factors such as testosterone are often presented as key causal factors in male violence. There are several ways to examine the role of testosterone in aggression. First is at the population or group level. The easiest groups to investigate are age classes. If testosterone is associated with aggressive behavior, we should see a relationship between rates of violence and aggression and ages in which testosterone is at its highest. We know that

within many populations, most homicides by males are committed soon after puberty, when men are in their twenties. Not surprisingly, this is when testosterone levels are at their peak.

We now know that there is significant variation among populations in the pattern of changes in testosterone levels over a man's lifetime. In western industrialized populations, testosterone peaks right after the age of twenty and declines by about 1 percent per year after the age of forty. Other populations, though, do not exhibit the same rate of decline, or in one case, no decline at all (Ellison et al. 2002). We have little or no data on homicide or aggression in these populations, but anecdotal information about the Ache of Paraguay suggests that these men are not very different, from the perspective of aggression studies, from American men. Until fairly recently, Ache men engaged in mass fights between rival groups armed with clubs, which often caused serious injuries. Moreover, I have witnessed acts of violence against women by Ache men. In one such case, a teenage girl was beaten with the blunt end of a machete for having run away from the village. This is not to say that the Ache are extremely violent. On the contrary, reconciliation is a central aspect of Ache culture, and conflicts rarely last very long. Nonetheless, Ache men retain the ability to be aggressive and violent despite exhibiting testosterone levels in the lower range of human variation. (For more about that variation, see Chapter 9.)

A second way to examine the relationship between testosterone and aggression is to look at variation in testosterone levels within a population. Again, one of the most straightforward methods is to compare the behavior of men of different ages in populations that show age-related declines or changes in testosterone. In boys in early adolescence, testosterone is not associated with physical aggression but has been reported to be associated with social dominance, that is, the ability to assert one's will over peers without resorting to physical confrontation (Schaal et al. 1996). The psychologist James Dabbs has found that male teens he classifies as delinquent exhibit higher mean levels of salivary testosterone. However, similar associations in older age classes have not been found (Dabbs and Dabbs 2000).

Some scientists have suggested that testosterone, in facilitating violent behavior, may act in conjunction with other factors associated with violence, such as alcohol consumption. Alcohol and other substances that suppress behavioral inhibition may act as releasers that allow testosterone to affect parts of the brain associated with aggressive behavior, such as the hypothalamus or the amygdala. David George and colleagues (2001) reported that men who were alcohol dependent and had a history of domestic violence had significantly higher testosterone levels than nondependent men with similar histories. But administration of exogenous testosterone, even in extremely high doses, has not been consistently found to be associated with violent behavior (Pope et al. 2000).

Adolescence is a turning point in male life history. Gone are the days of psychological and physical investment in growth with no competing interest in sex. With puberty comes the realization that life involves more than just playing, warding off boredom, and wondering what to eat. Sex and the accompanying costs now take center stage. However, the young men's trade-offs are about to shift once again. Soon the issue will be how much time and energy they should invest in finding and winning mates and how much in caring for offspring—and perhaps they will have enough resources left over to stay alive.

7

Sex and Fatherhood

ALLOW ME TO INTRODUCE the northern quoll *(Dasyurus hallucatus)*. The northern quoll is a small marsupial mammal, weighing in at about a kilogram. What makes this animal so remarkable is that, along with a few even smaller marsupial species, it exhibits an extremely rare trait for a mammal, a trait known as semelparity. That is they expend all their time and energy on one reproductive event and then die (Humphries and Stevens 2001). The vast majority of mammals are not semelparous but iteroparous, having multiple reproductive events and therefore being able to spread the costs over multiple bouts of effort. Semelparity is much more common in nonmammals such as insects and fish. The northern quoll, among mammals, is the poster child for reproductive effort.

Of all the concepts in life history theory, reproductive effort is among the most important, yet it is difficult to define. In essence, reproductive effort is the amount of time and energy devoted to tasks that result in more of one's genes being passed on to subsequent generations. We can narrow this a bit to "mating effort." But what is mating effort? In many organisms, mating behaviors are highly specific and easily identified.

For example, in deer, bucks challenge and fight other males to gain access to mating opportunities. Among primates, baboons are among many species in which males fight for rank position in the group, with higher rank leading to greater access to females. However, mating effort can also include providing food and other resources to mates and offspring. And in addition to behaviors, energy can be invested in building body tissues that lead to greater access to mates rather than in tissues that enhance survivorship.

Rising to the Occasion

Libido is obviously not unique to males. Indeed, the neuroendocrine and sex organ mechanisms involved are quite similar in males and females, having developed from common fetal structures. Conveniently, the development of male sexual motivation coincides with sexual maturation. This is largely the result of testosterone and other hormones. However, the relationship between testosterone and libido is subject to other important sources of variation such as cultural norms, overall health, and attitudes toward sex.

Among the most basic initial investments of mating effort is the development of sexual motivation and the physiology that allows males to mate with and inseminate females. The metabolic investment in this endeavor is probably very significant, and, interestingly, the physiology that has evolved for mating purposes can also serve as a direct cue of a male's intent.

Penile erections are a sign of sexual arousal in many vertebrates. In nonhuman primates, erections can reflect not only a state of arousal but also a more general interest in some social interaction. Thus erections often reveal a male's inner state, providing an honest signal of his intention or motivation. An anecdotal example of this comes from chimpanzees. A young male chimpanzee became interested in an estrous female and his penis became erect. However, when the group's alpha male suddenly appeared, the young lower-ranking male seemed to try to cover his erection with his hands, perhaps to keep the alpha male from viewing him as a rival and possibly attacking him (de Waal 2000). Although attributing intention to this act is speculative, this anecdote is an example of how the male erection is involved in aspects of social behavior other than mating. It also illustrates the way erections can indicate intent to invest time and energy in reproductive effort.

What actually happens, physiologically, to create an erection? In response to touch, sound, or visual stimulus, neurological and chemical signals involving nitric oxide are sent to smooth muscle tissue that surrounds the corpora cavernosa, spongy penile tissue that is responsible for creating and sustaining an erection. When these smooth muscles, which control blood circulation to the corpora cavernosa, are relaxed, blood is allowed to flow through the penis, causing the tissue to expand (Andersson and Wagner 1995).

The mechanics of penile arousal seem to be quite similar across mammalian species. Perhaps the primary difference between humans and other mammals is the decreased emphasis on olfactory cues in humans. Other mammals, including nonhuman primates, derive cues regarding female receptivity from scents that seem to stimulate male arousal. Despite the plethora of claims from cologne and perfume companies, there is little compelling evidence to suggest that human male sexual arousal is as sensitive to scent.

The initiation and maintenance of an erection is only one step in the path leading to fatherhood. Ejaculation is the process by which seminal fluid is ejected from the penis in response to sexual stimulation. Barring some problem with spermatogenesis, sperm within the seminal fluid makes its way through the vaginal canal with the potential end being fertilization of an ovum.

Ejaculation is a reflex-like action rooted in several neurological mechanisms that span from sensory nerve endings in the glans (the head of the penis), up the spinal cord, to the brain, and back to the smooth muscles around the seminal vesicles. Ejaculation can occur in response to arousal during sleep, in what are called nocturnal emissions or wet dreams, or in response to sexual stimulation. Nocturnal emissions are common during puberty and in adult men who experience long periods of sexual inactivity. Ejaculation through tactile stimulation involves mild friction to the penis, as is common during intercourse. During stimulation, the autonomic nervous system activates the sympathetic nerves to cause contraction of accessory sexual organs such as the vas deferens, prostate, and seminal vesicles. Semen, consisting of fluid from the prostate and seminal vesicles plus sperm, is mixed immediately before ejaculation. This fluid mixture is deposited in the posterior urethra, where it is stored momentarily, until contraction of the bulbocavernosus muscle ejects it through the penis via the urethra. Ejaculate commonly consists of one to five milliliters of fluid.

How does one get to the state of arousal that leads to sexual activity? Farmers, aware of the role of the testicles in affecting libido, long ago began to castrate animals to make them more manageable or to inhibit their reproduction. Some scientists took the idea that the testes were involved in libido to the point where they injected themselves with testicular tissue. Their reports suggested that some substance within the testes seemed to stimulate sexual arousal. These early studies, although flawed by the biased interpretations of their researcher-subjects, led to more research by endocrinologists. The eventual synthesis of testosterone in 1935 was a major breakthrough and sparked many decades of investigations into neuroendocrinology and behavior (David et al. 1935; Freeman et al. 2001; Hoberman and Yesalis 1995).

The study of human male libido by Alfred Kinsey and others, published in 1948, provided useful descriptions of variation in male sexual motivation. Little was known about testosterone outside the laboratory, however. In 1970 an unusual article appeared in the prestigious scientific journal *Nature*. The report was uncustomary for two reasons. First, it was published anonymously. Second, it was one of the first investigations to quantify male libido in field circumstances, outside the laboratory. The investigator reported associations between his own heterosexual activity and a male-specific characteristic that is linked with androgen production: beard growth. He lived and worked on an isolated island. On occasion he left the island and visited a woman with whom he had sexual relations. He reported that his beard growth, which he measured by carefully collecting and weighing his clippings, increased in anticipation of sexual contact with his partner. This was the first field report of a relationship between testosterone, libido, and a secondary sexual characteristic. At the risk of stating the obvious, it provided quantitative data that male physiology is strongly influenced by sexual motivation—that the arrow of causality between hormones and behavior goes in both directions.

A multitude of investigations since then have explored the links between testosterone and libido. Total absence of testosterone, due to injury, illness, genetic defect, or side effects of medication, commonly results in lower sexual motivation as measured by frequency of intercourse, masturbation, and ejaculation. But there is little evidence of any consistent dose-related effect between testosterone and sexual arousal (Anderson et al. 1992; Buena et al. 1993). In men with abnormally low testosterone (hypogonadal men) who are given testosterone treatment, measures of sexual arousal initially seem to respond in a dose-dependent manner, with higher doses yielding a stronger sex drive—but once testosterone reaches levels within the common range of variation in healthy men, higher doses of hormone cease to increase libido (Davidson et al. 1979). However, hypogonadism may be caused by factors such as injury, stress, illness, or a genetic disorder (Seftel 2006), any of which may influence the effect of testosterone treatment on libido. To get around these issues researchers can shut off the male hormone system with chemical blockers and replace testosterone in gradually increasing doses. Carrie Bagatell and colleagues (1994) chemically induced hypogonadism, restored testosterone at lower and higher doses, and found a dose-dependent relationship with libido similar to that reported by Davidson and colleagues. It is therefore clear that testosterone has a dose-dependent effect on men with very low testosterone levels and a negligible effect on men with normal levels.

A note of caution is warranted here. All the investigations of the relationship between testosterone and male libido have been conducted in western industrialized populations, and we now know that testosterone levels in men in industrialized societies are at the high end of the range of human variation (Bribiescas 2001a; Ellison et al. 2002). Emerging data on libido and erectile function from nonindustrialized populations suggest that there is much to learn (Gray and Campbell 2005).

But at What Cost?

Reproduction and survivorship are often competing needs in organisms. This is the basic trade-off discussed earlier: time and energy devoted to finding mates and producing progeny is often at odds with one's own survival. For example, in female lizards, survivorship is inversely related to the number of offspring produced per season. The mechanisms behind this relationship are many, including the inherent risks associated with gestation and protecting offspring, the energy required to produce and rear offspring, and changes in immune function that make females prone to various infections.

The trade-off between reproductive effort and survivorship in males has also been well documented in several species such as drosophila (fruit flies) and flatworms. The biologists Linda Partridge and Marion Farquhar illustrated this in a classic experiment. Male drosophila were housed with either virgin females (which are receptive to copulation) or nonvirgin females (which are not). The males that were able to mate exhibited significantly shorter life spans than males that could not mate (Partridge and Farquhar 1981). In flatworms, the process of spermatogenesis alone was enough to reduce survivorship compared with males that did not produce sperm (Van Voorhies 1992).

In a study of vertebrates, researchers surgically sterilized female brushtail possums *(Trichosurus vulpecula)* to prolong their period of receptivity to copulation. Males who were exposed to these females spent a longer time trying to mate than males exposed to unaltered females, and exhibited a significant decline in general health, potentially increasing their mortality (Ji et al. 2000). In primates, females of species that are monogamous tend to exhibit fewer white blood cells than females of promiscuous species—an indication of lower risk of infection associated with sexually transmitted diseases (Nunn et al. 2000). Limited evidence in support of a similar pattern is also available from human populations. In preindustrial Sweden, higher numbers of children born in a parent's lifetime were associated with higher mortality after the age of fifty in women, but not in men (Dribe 2004).

Human reproductive effort is also energetically costly (Ellison 2003). Sexual intercourse itself does not seem to have a significant impact on men's performance in physical exercise (Boone and Gilmore 1995), but there are other ways in which sexual activity can influence the expenditure of energy. The actual act of searching for women (assuming heterosexuality) takes a good deal of energy, and intercourse exposes men (as well as women) to risks that can require significant energetic expenditure. The assumption that men are likely to invest more energy in mating than women do arises from the notion that males have reason to seek more mating opportunities than females because the potential fitness benefits of additional mating are greater for males (Buss and Schmitt 1993). Granted, such patterns vary widely in contemporary societies, but in terms of our evolutionary history, this can be a reasonable assumption.

Attitudes toward sex-related risky behavior vary between cultures (Wight 1999). However, it is often the case that men and boys have fatalistic attitudes toward such risks. For example, heterosexual Nigerian men may see the risk of contracting AIDS as an inevitable peril associated with being a man and seeking sexual partners (Orubuloye and Oguntimehin 1999). Large, reliable data sets on sexual histories and risky behaviors are rare, but a remarkable survey conducted in Sweden provides some insights into the possible links between sexual behavior, risk, and detrimental outcomes. The researchers surveyed 814 boys regarding their age at first intercourse, sexual history, and certain risky behaviors. Significant relationships were evident between earlier ages at first intercourse (younger than fifteen) and smoking and drug use. Boys who had impregnated a girl were significantly more likely to have had more than ten sexual partners, used illicit drugs, engaged in binge drinking, and had sexual intercourse at least twice on first dates (Edgardh 2002).

Social factors make it hard to specify the association between risky behaviors, survivorship, and reproductive effort in human males. For example, social status and expectations of longevity probably contribute to the findings of the Swedish study. Nonetheless, we can view social status and risk assessment, not as confounding factors, but as contributory life history variables that influence trade-off decisions between survivorship and reproductive effort. Clearly, the questions are much more complex for human males than for drosophila or brushtail possums. It may be that reproductive effort can take on more forms in male humans than in males of other species. In the same way that somatic resources can be translated into and stored in extrasomatic forms, as when someone uses bodily energy to earn money, perhaps reproductive effort can be shifted between different social currencies such as status.

An interesting relationship between testosterone and metabolic costs has been found in the context of a category of behavior that is inherently linked with male mating effort: competition. When men are anticipating a competitive interaction, their salivary testosterone levels rise. Losers exhibit a decline in salivary testosterone, while winners maintain their enhanced precompetition levels. This finding is very robust, having been observed in men involved in physical competitions (Booth et al. 1989; Suay et al. 1999), in men competing in nonphysical contests such as chess games (Mazur 1992), as well as in spectators at sports competitions (Bernhardt et al. 1998). Such increases in testosterone before competition are not evident in women (Kivlighan et al. 2005).

The evolutionary implications of this competition-related rise in testosterone have been debated, mostly in relation to the possible effects of testosterone on behavior and the significance of population variation (Bribiescas 1998; Mazur and Booth 1998). The common assumption is that testosterone increases associated with competition reflect the evolution of a *behavioral* response to competition. This is somewhat odd since similar increases in testosterone are not associated with greater aggression or libido. So why does testosterone rise in response to competition? Perhaps the intended effects are not on behavior but on energy metabolism. Larry Tsai and Robert Sapolsky (1996) relate the effects of short-term testosterone increases, like those commonly found in competitive studies, to energy expenditure. In basic terms, Tsai and Sapolsky report that administering testosterone to muscle cells induces both brief (within two minutes) and extended (six hours) increases in sugar uptake and expenditure of energy. They suggest that these boosts in testosterone may prepare a male's body for a competitive interaction by increasing the amount of energy available to muscle tissue that may be called into action during competition.

As should be clear by now, the cost to humans of searching for and securing mates is very difficult to quantify. Indeed, it could be argued that the search for mates is a constant enterprise, interrupted only by the occasional cue that a receptive mate is at hand. Nonetheless, subtle pieces of evidence do suggest that the allocation of time to seeking and, perhaps, to guarding mates imposes significant costs on men. Evidence emerges from behavioral and physiological proxies related to mate seeking. Among the most salient physiological cues is testosterone, but the use of this hormone as a proxy demands caution. Individual variation in testosterone levels is not likely to be useful because of the vast range of variation in humans, but looking at testosterone variation between age classes may provide insight into the relationship between this hormone and behavior.

We might hypothesize that time allocated to risky behavior *related* to increasing access to mates is a cost. Examples might include consumption of alcohol and tobacco, driving recklessly, and other behaviors that reflect poor judgment. Whether any of these behaviors is actually associated with increased access to mates is an open question. But it is interesting to note that all of these factors are especially germane to adolescent and young adult males.

Among the Ache, success at foraging and hunting has been found to be related to increased reproductive fitness (Kaplan and Hill 1985). Therefore, it is relatively safe to assume that hunting is a decent proxy for the allocation of time to mate seeking in foraging societies. In contemporary societies, men's acquisition of wealth, prestige, or political power can be viewed as mating effort. However, categorizing any and all endeavors as mating effort would diminish the utility of the concept of reproductive effort.

Sperm: Quantity and Quality

The hallmarks of spermatogenesis are continuous production throughout life and a broad range of variation in production numbers among individuals. Indeed, sperm count has often been used as a measurement, albeit a questionable one, of male fertility. But sheer numbers are not the only important characteristic of sperm. The continuous production of sperm means that male germlines have a quality-control problem. Errors in gametes, unlike those in somatic cells, can result in aborted conceptions, flawed offspring, or defects that continue in the germline for generations. This is obviously not a preferred outcome: given the trade-off between reproductive effort and survivorship, if a male gains access to an opportunity to mate, the effort should not be wasted. Female germlines have similar quality-control issues, but the mechanisms are very different because ova are made while the female is still a fetus and stored in a state of suspension for the remainder of her life. So, for females, the challenge is more related to preservation of gametes.

Given the vast number of sperm that men produce, perhaps there is a range of tolerance for quality. It seems highly unlikely that each and every genetic defect (deleterious or neutral) is caught and repaired. But mechanisms are evident that seem to keep some semblance of control over genetic mistakes. Sperm DNA-repair mechanisms have evolved that not only identify potentially problematic sperm but act to rectify errors while maintaining enough mutations to allow adaptive selection if needed.

When DNA-repair mechanisms break down or become less than optimally efficient, male fertility may be compromised. How these mechanisms become less efficient—or how they work at all—is still under investigation. The sources of deleterious mutations are unclear, although certainly there are random mutations that can have significant effects on men's fertility. DNA errors on various chromosomes, such as on the Y chromosome, can also be a factor (Fox and Reijo Pera 2002).

While women usually produce one ovum per month, men make, on average, millions of sperm per day. If only one is necessary for fertilization, why produce so many? Also, do characteristics of sperm, such as count, concentration, motility, and morphology, contribute to variations in fertility? One would think that in an era of high-tech methods of reproductive medicine, these questions would have long been settled.

However, clinical research is awash in contradictory information regarding the effects of these sperm characteristics on male fertility.

Healthy men exhibit a wide range of sperm count and sperm morphology. Sperm counts from healthy men have been noted to range from one million to 120 million per milliliter of semen. Over the past century there has been significant disagreement about the minimal sperm count necessary for a man to be considered fully fertile. Minimal limits as defined by the medical profession have ranged from 10 million to 60 million per milliliter, although counts over 20 million are often considered optimal for full fertility (World Health Organization 1999). But why do sperm counts range so widely? The answer is unclear; activity, diet, environmental pathogens, and lifestyle have been suggested to be contributory (Bribiescas 2001a; Wong et al. 2003). Body and testicular temperature (Mieusset and Bujan 1995), nutritional factors (Wong et al. 2000), and frequency of ejaculation (Cooper et al. 1993) have also been suggested as mechanisms that may affect the production of sperm.

To determine the role of men in couples' fertility, aspects of sperm function such as number, concentration, motility, and morphology are commonly used to quantify male fertility. For example, David S. Guzick and colleagues (2001) collected semen samples from 765 men in infertile couples and 696 men in couples with known fertility. Their results are indicative of the broad range of variation and the overall normal distribution of variation. Sperm morphology was the most salient factor that discriminated between fertile and infertile men, but even in that characteristic, a significant overlap was evident between the two groups. Recent reports have provided further support to these findings (Nallella et al. 2006).

Sperm morphology and motility, but not sperm count, have been found to be positively associated with the success of intrauterine insemination (IUI), a procedure used to treat infertility (Montanaro Gauci et al. 2001). Similar results have been noted by other investigators (Hauser et al. 2001). However, still other researchers have found that men contributing to successful IUI had significantly higher sperm counts (Huang et al. 1996).

Worthy of mention in this discussion is the reported decline of sperm counts over the past century. Declines of 40–50 percent have been depicted in American men (Becker and Berhane 1997), and been attributed to various factors such as environmental pollutants (Bahadur et al. 1996). However, variation in sperm counts over time and among differing populations strongly suggests that caution is warranted in claiming such a decline. Further analysis has disclosed that sampling biases (Fisch and Goluboff, 1996; Saidi et al. 1999), changes in clinical norms (Bromwich et al. 1994), variation in quantitative methods (Emanuel et al. 1998), and unsuitable statistical analyses (Olsen et al. 1995; Safe 1995) may underlie many of the reported differences. Even if modest declines in sperm count have indeed occurred, there is no reason to believe that male fecundity has been compromised.

Men seem to be more buffered than women against the effects of caloric restriction or energetic output, though quantitative data from human males is extremely limited. Acute fasting does not affect spermatogenesis (Abbas and Basalamah 1986). Starvation has been found to have modest effects on the quality of human semen, but not enough to cause subfertility (Hulme 1951). Differences in diet, such as vegetarian versus nonvegetarian, are not associated with differences in sperm physiology (Jathar et al. 1976), although reproductive hormones may differ between vegetarian and nonvegetarian men (Howie and Schultz 1985). Somewhat independent of caloric intake is the impact of micronutrients. Of particular importance is zinc. Acute zinc deficiency compromises spermatogenesis and testosterone production (Chia et al. 2000; Fuse et al. 1999).

Energetic expenditure such as exercise can suppress sperm production, but observed declines are still well within the range of accepted fertility (Hall et al. 1999). In one study, for example, when male runners doubled their weekly running regimen, their testosterone levels declined and their cortisol levels rose. Sperm counts also fell but, again, remained in the normal range (Roberts et al. 1993). Comparisons between sedentary men and marathon runners revealed no significant difference in sperm count, morphology, or motility (Bagatell and Bremner 1990).

Our understanding of the range of human variation in sperm function is somewhat limited, but it appears that men in nonwestern populations tend to have lower sperm counts than their American counterparts (Fisch and Goluboff 1996; Fisch et al. 1996), but not so low as to represent a compromised ability to father children. Long-term studies of sperm count in nonwestern men do not exhibit any significant changes over the past few years. For example, in a study of Venezuelan men, sperm counts were well within the range of clinical western standards, and little or no change was observed over a span of fifteen years (Tortolero et al. 1999). A ten-year investigation of sperm count and quality in Korean men produced comparable results, although 19 percent of the subjects were diagnosed with azoospermia—that is, their semen contained no measurable amount of sperm—suggesting that the physiology of spermatogenesis in various parts of the globe merits further investigation (Seo et al. 2000). Similar studies of European populations also found no significant long-term changes in sperm count or quality (Andolz et al. 1999; Berling and Wolner-Hanssen 1997; Rasmussen et al. 1997).

Fatherly Investment

From a life history perspective, fatherhood involves trade-offs between investments in survivorship, reproductive effort, and offspring. These decisions about how to allocate time and energy involve a multitude of factors. First, the man must decide whether to invest time and energy in offspring he thinks are his. He must also decide whether to invest in the mother of his children. Mothers and children also have their own agendas that affect male reproductive strategies. Having a father around can be a crucial advantage if he provides food and helps with childcare. However, fathers can also be a liability. Men can be abusive to both mother and children, or they can simply be another mouth to feed and a drain on the family's resources (Hewlett 1992).

Let's diverge from the discussion of humans for a moment and explore a species in which fatherhood and high paternal investment are commonplace and seem to have been subject to high degrees of positive selection. Perhaps the best fathers in the natural world are emperor penguins *(Aptenodytes forsteri)*. After laying a single egg, the mother penguin leaves the nest to feed for several weeks. During her absence, the father stands almost motionless, protecting the egg from the ice and the elements by tucking it under his belly feathers and balancing it on his feet. The father is totally committed to the egg and does not feed until the chick is hatched. When the survival of offspring is highly dependent on paternal investment and mating opportunities are limited, males tend to be "good" fathers.

The penguin example, however, is not totally applicable to mammals.

In birds, external gestation creates the possibility of direct prenatal male investment and care. This is not the case in organisms with internal gestation. Nonetheless, high paternal investment is evident in some primates. Tamarins, for example, exhibit relatively high paternal investment. Tamarin females commonly give birth to large-bodied twins that need considerable parental care after birth. Tamarin males who are candidate fathers (tamarins are polyandrous, with several males mating with a given female) aid in caring for the offspring by carrying them on their backs, allowing the mother to feed more efficiently and perhaps become sexually receptive and fertile sooner as the result of lower energetic expenditure. Evidence from tamarins also suggests that finding mating opportunities and establishing territories may be dicey endeavors for young males. On occasion, young males stay with their natal group and invest in sibling care. This "helper at the nest" behavior is not uncommon in species with strong competition among males for mating opportunities (Goldizen 1987).

It may be useful to hypothesize that human males are influenced by similar challenges to optimize the fertility of a particular mate, the survival of offspring, and the possibility of seeking other mating opportunities. Kim Hill and Magdalena Hurtado (1992) investigated the relationship between mating opportunities and the development of fathering strategies in the Ache of Paraguay. First they needed to determine if the presence of a father had any effect on the well-being of offspring. To devise a quantitative measure of such effects, they examined the impact of paternal survivorship on child mortality. They found that among the Ache the children of fathers with higher survivorship had lower mortality: having a father around improved a child's chances of surviving into adolescence. The effect was evident from soon after birth until the age of fifteen, when Ache children had become relatively independent. A similar pattern was evident when the father's absence was caused by divorce rather than death.

The effects of the availability of resources on trade-offs between investment in offspring and in mating effort are difficult to measure; the relevant resources vary widely, and mating effort is not easy to quantify. Nonetheless, Ache women, who often have several sexual partners, tend to identify men who are successful hunters, and thus presumably able to provide meat for their families, as probable fathers of their children more often than expected by chance, perhaps reflecting the role of resource availability in generating mating opportunities for Ache men (Kaplan and Hill 1985). In more modernized populations, income and education may be proxies for a man's ability to provide resources. Among rural men in Belize, income and education were associated with lifetime number of sexual partners (Waynforth et al. 1998). Ruth Mace (1996) reported that, among Gabbra pastoralists of Kenya, wealth as measured by camel herd size was positively associated with number of children in both men and women, although the association was more pronounced in men.

The anthropologist Mark Flinn, who has studied the role of fathers in a series of investigations on the island of Dominica, has found that fathers interact with children substantially less than mothers do, but that the paternal interactions are still significant. In early childhood (between birth and four years of age), approximately 10 percent of children's social interactions involve their fathers. About 44 percent of their interactions are with their mothers. The remaining interactions involve grandparents (about 18 percent), siblings (16 percent), more distant relatives, and people outside their families.

Simple logistical factors also affect the extent to which fathers invest in or interact with offspring. For example, not surprisingly, Flinn found that fathers who live with their children interact with them more than fathers who live elsewhere. He also noted that fathers tend to interact more with sons than with daughters, and with genetic offspring than with stepchildren. The father's mating status can also affect the amount of time spent with offspring. Fathers who were single spent more time with their children than those who had acquired a new mate (Flinn 1992; Flinn and England 1997).

There is a robust literature on the role of hormones, particularly testosterone, in the development and maintenance of paternal behaviors in nonhuman organisms. Researchers at Harvard University, using measurements of salivary testosterone, have found evidence that men in committed relationships exhibit lower testosterone than men who are not in steady relationships (Burnham et al. 2003; Gray et al. 2002). To discover whether this finding was applicable to other cultural settings, the anthropologist Peter Gray (2003) looked at salivary testosterone in polygynous and monogamous men, as well as unmarried men, in Kenya. His hypothesis was that polygynous men would have lower testosterone levels. His rationale was that demographic data showed that polygynous men had more children and were less eager to seek additional mates. Gray interpreted this as greater parental and less mating effort. He further predicted, on the basis of results from the United States (Gray et al. 2002), that married men would have lower testosterone than unmarried men because of their investment in parenting over mate seeking. The results did not support either hypothesis. Polygynous men in fact had higher testosterone than monogamous men, a possible representation of the greater mating effort evidenced by their acquisition of multiple wives. There were no significant differences in testosterone levels between married and unmarried men. But in a subsequent study in urban China, Gray found that fathers had lower testosterone levels than unmarried men and married men without children. Clearly more research is needed to complete this picture.

Paternal Uncertainty

Whereas female fertility is limited by the amount of energy it takes to produce offspring, male fertility is limited by the availability of mates. Granted, most men are highly unlikely to find themselves in situations in which mates are unlimited. Nonetheless, the effects of the availability of mates on male reproduction are clear under unusual circumstances. For example, the world record for producing children by a woman is 69 by a Russian mother who had a series of multiple births. In contrast, a sultan of Morocco produced more than 800 children, according to dependable records. Less verifiable reports of Chinese emperors and other royal personages suggest instances of individual men having fathered thousands of children (Guinness Media 1996; Daly and Wilson 1983). Regardless of royal status, an average man has enough sperm per ejaculation to repopulate a good portion of North America.

With such potentially large payoffs in lifetime fitness, men can reap huge fitness benefits by acquiring more chances to impregnate women. However, in the real world, a man's mating opportunities are limited by circumstances such as women's preferences and his own attractiveness as well as by the sheer number of available mates. In addition, lifetime reproductive success is a function not only of mating opportunities but also of offspring survivorship. A man might father a thousand children without any gains to his fitness if they all died without giving him grandchildren.

Men can contribute to children's survival by investing time and energy in providing food for them or in protecting them from danger. So why is it that women tend to invest more time and energy in offspring than men do? The answer lies in the simple factor of parental identification. A woman knows who her offspring are because (except in a few cases of surrogate motherhood) they emerge from her own body. However, because of internal fertilization, men cannot be certain of the paternity of the children they believe are their own. Therefore, in the evolution of our species there would have been less selection for males to invest in offspring, because the offspring might be carrying some other man's genetic inheritance.

Uncertainty about paternity has led to mate-guarding strategies in many species, and has played a very important role in the evolution of paternal investment in offspring. In this uncertainty we can begin to see the reasons behind the evolutionary emergence of apparently dubious traits such as the absurdly large claw of male fiddler crabs, the willingness of male baboons to fight one another despite the risk of serious injury, and the propensity of male humans to take dangerous chances in the name of attracting females. Male humans, like other male mammals, have been shaped by uncertainty about paternity and the risk of being cuckolded, which make it more beneficial for them to put their efforts into acquiring more mates than to invest in fewer mates and the survival of offspring.

A traditional Mexican joke goes as follows. The father of a family was cursed to die at a predetermined hour by a *bruja,* or witch. The appointed time came and, to his relief, the man of the house remained in excellent health. His good cheer was short lived, however. A loud thump at the door revealed that the milkman had dropped dead. Thanks to the evolution of internal fertilization, men can never be absolutely sure that children they think they fathered are genetically theirs. The impact of this uncertainty on the evolution of various species is only now being unveiled. For example, scientists used to assume that, in hierarchically organized mammalian groups such as those of seals and macaques, the dominant male fathered most of the offspring. But genetic paternity tests have consistently revealed that alternative mating strategies relying on female choice and "sneaky" copulations result in more offspring than previously suspected (Amos et al. 1993; Berard et al. 1994). The impact of female choice on paternity can vary with ecological circumstances. For example, some seals copulate on beaches and some do their mating at sea. In species that mate on land, the females are often subject to mate-guarding strategies by large, aggressive males, with the result that a small number of males father the majority of offspring each season (Le Boeuf 1974). However, in species that copulate at sea, in which females have greater mobility and access to resources, paternity is more dispersed (Coltman et al. 1998).

Paternity is more difficult to study in humans, but we know that in some societies paternal uncertainty is an accepted fact of life and is even quantified and incorporated into their definitions of fatherhood. Among the Ache, for example, men and women often have multiple sexual partners in early adulthood, becoming mostly monogamous after the age of fifty or so. Consequently, attribution of paternity is sometimes uncertain. Fatherhood is assigned according to a man's probability of being the genetic father. The Ache are well aware of the association between intercourse and conception. And they understand that if a woman has sexual relations with more than one man during the time span in which conception is believed to have occurred, there will be some ambiguity about her child's paternity. Accordingly, the Ache assign a child both a primary father and secondary fathers (Hill and Hurtado 1996). A primary father is the one who had most frequent intercourse with the mother in the month preceding her first missed period; secondary fathers had more limited sexual access. While biological factors determine genetic paternity, defining fatherhood is a complex process in many societies. For example, Hill and Hurtado caution that among the Ache, as with many other tribal populations, the term for father also refers to brothers of the father and may not indicate any paternal role.

Decisions, Decisions

Once a man has decided that a certain child is his, how does he determine how much time and energy to devote to that child? A man cannot be in two places at once. Contemporary fathers, along with mothers, must cope with the challenge of balancing family life with other demands such as careers and the everyday chores such as laundry and lawn mowing. Both men and women throughout evolutionary history have had to choose between investing in current offspring and pursuing further opportunities to mate. Traditionally, this has been viewed as a predominantly male issue. However, evolutionary anthropologists have questioned this assumption and have demonstrated quite convincingly that women face similar trade-offs. Sometimes it is evolutionarily beneficial to abandon offspring or rely on other care providers in order to gain access to additional mating opportunities (Hrdy 1981, 1999). But the costs and benefits of these decisions are different for men and women. For a man, abandoning offspring would probably result in increased morbidity and mortality among the offspring. But there is a chance that the children are not his, and if they are not, his fitness cost of abandoning them is virtually zero. For a woman, in contrast, abandoning her offspring may result in greater future reproduction, increased investment in other children, or better access to high-quality mates with good genes, abundant resources, or both. But her fitness will definitely take a serious hit, since she has already invested at least nine months of time and energy in producing the offspring.

Men have a finite amount of time to find mates and father offspring while dealing with other men and keeping up their own physical well-being. Using hunter-gatherer populations as a model of human evolution, we can say that decisions about foraging were probably a pivotal aspect of everyday life for our hominid ancestors. When does one forage, for how long, and for what specific resources? Research among contemporary foraging groups such as the Ache, the Hadza of Tanzania (Marlowe 1999a, 1999b), and Australian aboriginal groups has illuminated the importance of time allocation to men (Bliege Bird 1999). Not only must men allocate time between foraging and other activities, but the time spent foraging is subject to decisions about resource choice. Ache men tend to prefer hunting wild game over gathering edible plants, even though the caloric return would sometimes be higher if they concentrated on the plants. This departure from optimal foraging may indicate the high social value of hunting in addition to the caloric payoffs (Hawkes and Bird 2002). In addition, men spend much of their time building and maintaining alliances and relationships with other men (Bailey and Aunger 1989). Observations of men in foraging populations, as well as males of many other primate species such as baboons and chimpanzees, have shown that time devoted to relationships with other males is an important aspect of male life.

Clearly, men's (and women's) time-allocation strategies will differ depending on their respective ecologies. For example, the time needed to acquire food will be different for those living in foraging societies than for those living in cities. An investigation conducted among the Huli of

Papua New Guinea compared the time allocation of rural dwellers with that of Huli people who had migrated to Port Moresby, the nation's capital. It revealed that men spent more time on subsistence activities in rural settings than in the city. In addition, in the country women worked significantly more hours than men did, but this discrepancy disappeared in the urban setting (Umezaki et al. 2002).

Traditionally, paternal care has been envisioned as a way of increasing the survivorship of a man's genetically related offspring. However, since internal fertilization always results in uncertainty about paternity, paternal investment in children may always be a form of mating effort, a means of gaining access to their mother (Hawkes et al. 1995). To test this hypothesis, Frank Marlowe (1999b) compared the amounts of time Hadza forager men invested in genetic children and stepchildren. His prediction was that if paternal investment was solely related to mating effort there should be no difference between men's investment in step- and genetically related children. He found, however, that stepchildren received less care than genetic children, a result supporting the notion that paternal investment is not merely a form of mating effort. A study of American men found that they tended to spend more time per week with the offspring of their present relationship than with children of previous relationships or stepchildren, even after controlling for the confounding effects of differing residences (Anderson and Kaplan 1999). And an investigation of a similar idea among British men indicated that their degree of investment in children correlated with their assessments of their wives' fidelity and the children's resemblance to themselves, implying that men will preferentially invest in offspring that are more likely to be their progeny (Apicella and Marlowe 2004).

Few studies have been successful at mapping life history concepts onto modern-day situations involving monetary capital and contemporary reflections of paternal investment. The study of American men mentioned above did address such factors, assessing paternal investment using four variables: probability of child attending college, probability of child attending college and receiving funds from father, current financial expenditures on children, and weekly amount of time spent with children between the ages of five and twelve. The findings provide evidence that contemporary American fathers invest more in genetically related children than in stepchildren, and that they invest more in children from their present relationship with whom they share a residence. Residence seems to be an important factor in male parental investment. It may be that, in human evolution, shared residence has suggested a greater chance of paternity.

What can be certain is that men make unconscious evolutionary choices every day involving the amount of time and resources to invest in children. Some men opt out of this type of investment altogether, choosing instead to mate with many women and produce many offspring. Some are more monogamous and invest heavily in children. Why the differences? Surely contemporary conditions of morality, religion, and social pressure have some influence on this choice. However, from a life history perspective, it may be worth considering other factors such as extrinsic mortality. Men who live in an environment in which high mortality from accidents or violence is the norm may be more likely to invest in mating opportunities than in children, since acquiring more mates might yield a greater fitness payoff than investing in young who would face the same dangerous environment. This is not to suggest that such life history decisions are made consciously, but it is worth considering whether selection has favored complex neuropsychological mechanisms that would predispose men to make certain decisions.

8

The Male Furnace

MEN AND WOMEN HAVE DIFFERENT needs for energy, and different strategies for managing it. Metabolically, for example, men do not invest nearly as much in reproduction as women do. Men do not menstruate, gestate, give birth, or (normally) breastfeed babies. Also, male bodies are more metabolically expensive than female bodies because of differences in tissue composition. Male mammals have been designed by natural selection to burn energy at a higher rate than nonpregnant females. In many ways, these departures from female energetic needs shape mammalian male life histories.

Parental Contributions

Producing offspring requires converting energy harvested from the environment into viable, healthy individuals capable of passing their genes to subsequent generations. In many organisms such as bacteria, this conversion is readily evident: the size of daughter cells is approximately half that of the parent cell. Among some insects, the amount of energy needed to produce offspring is proportional to the amount of energy stored in the biomass of the offspring. For example, scorpion fly males provide food in the form of insect prey to females in order to acquire mating opportunities; the size and quantity of the prey provided often predicts the size and quantity of the eggs laid by the female (Thornhill 1976). In some mammals, such as elephant seals, there is a strong correlation between nursing, maternal weight loss, and calf weight gain, illustrating the direct transfer of energy from mother to offspring (Carlini et al. 1997). Males do not convey energy to their offspring in the same manner.

The contrasting metabolic investment of men and women in reproduction begins with the size of gametes. With the evolution of sexual reproduction came the emergence of anisogamy, in which male and female gametes are of different sizes. In virtually all sexually reproducing species, ova are much larger than sperm (Bell 1978). But men and women also invest different amounts of metabolic energy after conception and after birth. Internal gestation is also one of the primary contributors to differences in reproductive investment. Since these traits became established within the mammalian lineage, females have borne the bulk of the costs of producing offspring. This inequity has many behavioral and life history ramifications. Evolutionary psychologists have suggested that jealousy evolved as a result of unconscious fear of being left with the reproductive check (Buss et al. 1992). Females do not want to be abandoned with high-maintenance offspring, especially after devoting so much time and energy to their development, and males do not want to be cuckolded and find themselves providing for offspring that are not their own.

Are there species in which the costs are more equitable or even switched between males and females? Absolutely. In many avian species males and females share equally in the care of offspring, usually by sitting on the clutch of eggs or by foraging for food to bring back to fledglings at the nest. These are tasks that can be done by either males or females since gestation is external. In a few species of mammals, natural selection has favored the evolution of offspring that require a tremendous amount of postnatal care, even by human standards. Some South American primates, such as the tamarins mentioned in the previous chapter, are among these rare mammals. They are among the smallest of primates, weighing just a few pounds, but mothers ordinarily produce twins. Moreover, these twins are exceptionally large, making up about 20 percent of the mother's mass. That would be the equivalent of a 120-pound woman bearing a 25-pound infant. A mother tamarin would be hard pressed to care for such large and demanding progeny on her own.

Consequently, these primates have evolved a social system that commonly includes polyandry and significant investment of time and energy by males. The mother's partner males carry the infants on their backs as they make their way about the neotropical canopy. In captivity, the effects of paternal investment are strikingly evident in male cotton-top tamarins *(Saguinus oedipus),* which lose 1–10 percent of their body weight in the first twelve weeks after the infants' birth (Achenbach and Snowdon 2002). In the wild, infant carrying is a central aspect of male caregiving, resulting in a 21 percent increase in daily energetic expenditure by males (Tardif 1997).

Why would male tamarins and marmosets be part of a system of polyandry and extensive paternal care if mammalian males are selected to mate with many females? The answer may lie in their size. As the smallest of primates, they may be subject to greater predation pressure than your average primate, in danger of becoming a meal for harpy eagles and cats such as jaguars and ocelots. Therefore, it may be a good idea for them to produce two offspring at a time in case predation catches up with one. What about their infants' large size? At this point we can only speculate, but it may be that selection has not affected fetal size—that, instead, adult size has decreased during these species' evolution. Why might the adults have shrunk? We're not sure. Perhaps as an adaptation that made them more adept in the higher, thinner branches of the canopy, or perhaps as an adaptation in response to becoming an insect predator. But this doesn't address the unusual role of males. Perhaps infant carrying by the male provides protection of the offspring, presumably his, from predators. In addition, freeing the mother from the drain on her energy of carrying such a large load may allow her to come into estrus sooner. We know that stress due to low caloric intake, high energy expenditure, or both can delay the resumption of ovulation. By shouldering (literally) some of the burden, the male may be improving his chances of fathering more offspring sooner rather than later.

There are no mammals in which males' metabolic contribution to the gamete or gestational stage of reproduction is greater than that of females. This does, however, occur in invertebrates. Male insects in such species as Mormon crickets and butterflies produce packets of gametes known as spermatophores. These are relatively large packets of sperm and liquid nutrients that the male transfers to the female's body during mating. Females use these packets to fertilize their eggs, but also for food. Consequently, in these species males are often the choosy sex, with a male evaluating the advances of many females before choosing one he deems worthy of his major metabolic investment (Boggs and Gilbert 1979; Gwynne 1981).

An interesting variation on this theme was proposed by the psychologist Donald Dewsbury. He suggested that mammalian males may exhibit some level of mate choosiness because of the limitations on how often they can ejaculate. Nonseasonal male mammals produce sperm more or less continuously and constantly, but the sperm are delivered to females in large quantities at discrete intervals, when a male ejaculates. Dewsbury argued that the delivery system can outpace the production of sperm, at least in the short term, resulting in constraints on the frequency and quality of ejaculations. Consequently, mammalian males may exhibit some level of choosiness because of this potential constraint (Dewsbury 1982).

This idea has some merit in regard to male humans given the time it takes to regain the ability to ejaculate after an orgasm, commonly referred to as the refractory period. The human refractory period can range from several minutes to hours, depending on age, stress, and general physical condition (Aversa et al. 2000). Nonetheless, there is little evidence to suggest that sperm becomes significantly depleted after each ejaculation. Sperm counts do decline, but not to the point where a male would be considered infertile (Nnatu et al. 1991; Oldereid et al. 1984). The refractory period seems to be a time of recharging the nerves and muscles involved in orgasm and ejaculation. But why has such a mechanism evolved? One would assume that males with shorter refractory periods would have an advantage over their competitors. Imagine a scenario in our evolutionary past where two or more males are competing for access to receptive females. A male with a short refractory period would have more potential opportunities to inseminate these females than a rival who had to wait around for things to perk up on his end. While the evolutionary origins of refractory periods are open to debate, it is clear that refractory periods are not static and can adjust in response to mating conditions.

In addition to youth and good physical condition, variation in sexual partners can significantly shorten the time between ejaculations. The phenomenon has been dubbed the Coolidge Effect, after the former U.S. president. The myth goes as follows. President and Mrs. Coolidge were visiting a chicken farm. As the host was showing them around, the first lady asked if the rooster was available to service the hens at any time. The farmer, proud of the quality of his poultry, proclaimed, "Absolutely!" Mrs. Coolidge, the story goes, asked the farmer to bring this to the attention of the president. When he did so, Coolidge asked if the rooster received a different hen every night, to which the farmer responded affirmatively. "Make sure you tell that to Mrs. Coolidge," commented the president.

The Coolidge Effect has been noted in many mammals. Among sheep, rams that were given opportunities to mate with novel females were able to ejaculate almost immediately after presentation of each female, up to more than half a dozen. Rams who were offered consecutive mating opportunities with the same female, in contrast, exhibited a rapid and sustained lengthening of refractory periods.

Male choosiness also occurs in species in which males bear most of the energetic burden of gestation. Probably the best-known example is the sea horse. The female sea horse's eggs are deposited in a pouch located on the male's belly. The male fertilizes the eggs with his sperm and carries the offspring to term. The metabolic burden associated with male gestation makes the males show particular care in selecting mates (Berglund and Rosenqvist 2003).

After the offspring are born, the primary difference in metabolic investment by parents is lactation. Among humans, the length of time women nurse their infants varies widely. Women exhibit several adaptive mechanisms that regulate the need to invest energy in nursing a current child and the reproductive benefits of conceiving future children. The hormone responsible for producing mother's milk, prolactin, suppresses ovulatory function at the hypothalamus. As long as an infant is suckling, tactile nerve connections that go directly to the hypothalamus from the nipple stimulate the production of prolactin and milk and make ovulation less likely to occur.

The anthropologists Melvin Connor and Carol Worthman noted that !Kung hunter-gatherer women of the Kalahari Desert had some of the longest interbirth intervals in any human population. Their observations of the women's nursing patterns, as well as hormonal data, indicated that frequency of nursing was an important contributory factor to variation in interbirth intervals and differences in population fertility (Konner and Worthman 1980). For many years, subsequent data from other populations supported their conclusions. However, even when nursing frequency is taken into consideration, there is still significant variation in the time it takes for women to resume ovulation after childbirth. In addition to suckling patterns and intensity, the mother's health and condition may play an important role in the resumption of ovulation (Valeggia and Ellison 2001).

But what if men were the ones who nursed babies? If men lactated, women would be able to resume ovulation sooner, thereby allowing men and women to produce more children. This question may seem to have an obvious answer: although men have breasts and nipples, they can't lactate. Wrong. With appropriate hormonal stimulation, breast growth is quite easy to achieve. The successful treatment of transsexual men during their change to female status bears witness to this. Furthermore, men can produce breast milk. Occasionally this occurs when men are under severe stress or develop prolactin-secreting tumors (De Rosa et al. 2003; Sonino et al. 2004). Along with breast growth and shrinkage of the male genitals, prolactin-secreting tumors stimulate milk production. It would seem relatively simple for natural selection to develop the ability to produce prolactin and its supporting hormones in men.

In 1994 Charles Francis and colleagues reported the discovery of lactating males in a species of Malaysian fruit bat *(Dyacopterus spadiceus)*. They described capturing ten males, all with functioning lactating breasts. This *may* be the first known case of naturally occurring male lactation in a wild, free-ranging mammalian species, although the researchers note that other factors such as phytoestrogens or liver dysfunction may stimulate breast development (Francis et al. 1994). Further studies are necessary to confirm this phenomenon. Nonetheless, if male lactation in this species is indeed the result of natural evolutionary processes, it is clear that several conditions must be present in order to make it adaptive.

First, males would need to develop a mechanism that would shield their reproductive function from the inhibitory effects of prolactin. Prolactin causes reproductive suppression in males in the same manner that it inhibits ovulation in women. Prolactin blocks the production of GnRH by the hypothalamus, the secretion of gonadotropins such as FSH and LH, and ultimately the release of testosterone. However, males may tolerate reproductive suppression if the potential reproductive payoff, perhaps in increased offspring survivorship, is higher than the potential benefits of maintaining sexual function. In addition, if a species is monogamous and reproduction is limited by the female partner's ability to conceive and bear young, then it will make sense for males to be involved in offspring care even to the point of lactating, especially if this allows the females to resume ovulation sooner (G. Bentley, pers. comm). However, a lactating male might not be able to mate or produce sufficient sperm because of the effects of prolactin, in which case having the female ovulate sooner would be a moot point. Perhaps confidence about paternity would have to be extremely high to merit such high investment by a male.

It is unclear whether these conditions are evident in the Malaysian fruit bat, but without them evolution of male lactation would be highly unlikely. As for humans, the emerging picture from comparative, theoretical, and medical research is that direct metabolic investment in offspring by the male reproductive system has been minimal compared with that of the female reproductive system. This is not to say that males do not invest a significant amount of energy in reproduction. Male energetic investment in reproductive effort manifests itself in other ways, such as investment in somatic tissues that may augment their physical competitiveness with other males or perhaps enhance their attractiveness to females.

Energy

The term "energy" keeps coming up in our discussions. But how is energy measured? Physiologists quantify the use of energy in calories per unit of time, or metabolic rate, typically described as the number of calories expended per day. Energy use is broken down into more specific components: resting metabolic rate (RMR), that is, energy used during periods of inactivity, a measure of the minimal amount of energy required to stay alive; sleeping metabolic rate (SMR); and basal metabolic rate (BMR), the averaged use of energy over the course of a day's general activity. I will use BMR since it most accurately reflects daily life. Related terms include "catabolism" and "anabolism." Catabolism is the active breakdown of tissues, often for the purpose of liberating stored energy. The liberation of calories from fat stored in adipose cells during energetic deficits is an example of a catabolic process. Anabolism is the creation of metabolically active tissue, such as the growth of muscles in response to weightlifting or other exercise.

Human males metabolize, catabolize, and anabolize energy quite differently from human females. Distinct sex-specific patterns of energy use form the underpinning of the sex differences in reproduction, growth, and senescence. Moreover, the processes of metabolism, catabolism, and anabolism are not random. They occur under specific circumstances when energetic demands merit a certain response. Obviously, given the importance of energy management, the regulatory mechanisms must have been subject to natural selection. Therefore, it is logical to assume that individuals with heritable mechanisms that allocate energy and manage these processes in the most efficient manner would have a selective advantage over less efficient individuals. This is one of the most important concepts in life history theory.

Not surprisingly, larger organisms use more energy than smaller ones. But this changes when we begin to talk about *relative* metabolic needs, that is, the amount of energy needed per unit of mass. We also need to consider whether the organism is a homeotherm, generating its own body heat as a mammal does, or an endotherm, depending on environmental sources of heat as reptiles do. For now, we'll confine our discussion to mammals. As homeotherms, mammals are dependent on internal metabolism to burn fuel to maintain body temperature at an optimal level. Elephants have higher absolute energy needs than mice, but mice burn more energy per unit of mass. Because of their small size and large surface-to-volume ratio, mice lose heat faster than elephants and must maintain a higher rate of energy metabolism. In humans and other mammals, there are also other significant sex differences in BMR. On average, males have higher metabolic rates than females, meaning that per unit of time males burn more fuel just standing around than females do. Why the difference?

First, men have higher absolute energetic demands simply because men, on average, are larger than women. But when body size is taken into consideration, male metabolic rates are still higher. Cellular differences in BMR or differences in the way male and female tissues burn energy are key. Why? The answer can be found in differences in body composition between men and women. Different tissues have different requirements for calories, micronutrients, and even oxygen. For example, brain tissue is very picky. It only burns glucose, no carbohydrates, no protein, just sugar please—except under the most dire circumstances when it resorts to much less optimal sources of energy such as ketone bodies. In addition, brain tissue needs a constant, uninterrupted supply of oxygen. More than a few minutes of deprivation results in acute cell death and irreversible brain damage. Indeed, the brain is, for humans, the most demanding mass of tissue. It accounts for about 20 percent of the basal metabolic rate, twenty-four hours a day, seven days a week. No exceptions, no negotiation. This in itself is an interesting aspect of human evolution. What selective forces were at play to evolve such a demanding organ? This question would fill a book of its own.

On average, men carry more muscle mass and less fat than women. Men have a greater proclivity to build muscle mass than women. Testosterone and other anabolic hormones are central mechanisms responsible for these differences. Men, on average, have about ten times more testosterone in circulation than women and therefore can generate greater anabolic responses to activity. In addition, men have more testosterone receptors in various parts of the body than women. But before we delve into muscle, we need to cover other tissues. Let's chat about fat.

Fat

Calorically, fat is a rich resource, if you can invest the small amount of energy needed to liberate those calories. Among forager groups, fat is gold. An experience I had with the Ache illustrates this quite well. After a hunt, since I had my trusty Swiss Army knife, I was asked to carve up a cooked carcass fresh from the pot. I think the animal on the menu was a coati—I'm not sure. Anyhow, when the carcass was handed to me, it was scaldingly hot. In trying my best to cut it up without suffering third-degree burns, I managed to drip most of the precious grease onto the dirt. The Ache were not pleased. Fat is important. If you can accumulate it on your own body (within reason), all the better.

Bodily fat is a crucial energetic currency that is central to many metabolic processes. In tissue form, fat is dubbed adipose tissue, and the storage cells are called adipocytes. Measuring a person's adiposity is somewhat problematic. Most fat is stored in peripheral subcutaneous tissues in the form of white adipose tissue. Another form of fat, which is stored in the trunk of the body around the internal organs and in the thighs of women, is brown adipose tissue, a form involved in the generation of body heat. Adiposity can be assessed by a variety of methods including measuring skinfold thickness, running a mild electrical current through the body and measuring resistance (bioelectric impedance), and measuring the circumference of certain bodily locations. In hospitals, adiposity is usually measured using dual-energy X-ray absorptiometry or DEXA, the gold standard for assessing body composition. As noted earlier, human males are leaner than females, beginning in adolescence and continuing throughout life. The source of this difference between the sexes lies in greater fat catabolism as a result of testosterone in men and greater efficiency in fat deposition in females as a result of estrogens.

If adipose tissue is so important, how does the brain know how much of it is available? What serves as the fat accountant? Here we turn again to endocrinology. Various hormones are involved in fat metabolism and energy management: insulin, ghrelin, adiponectin, and numerous other agents. But one hormone—leptin—stands out as the "fat reporter" to the brain.

Leptin is a protein hormone that is predominantly secreted by fat cells (Ma et al. 1996; Maffei et al. 1995), although it is produced in much smaller quantities elsewhere in the body. Initial studies suggested that leptin is a fat signaler to the hypothalamus, conveying signals that indicate available energetic stores. In other words, leptin is the bank statement that tells the hypothalamus how much capital is stashed away in the fat vaults. Leptin has been shown to be involved in many aspects of energy balance including feeding and hunger. Genetically altered rats that lack the leptin receptor eat uncontrollably, presumably because the hypothalamus fails to detect signals that fat stores are available. Sensing no leptin, the brain believes the body is starving and demands incessant eating. Consequently, rats without leptin receptors become quite obese. This finding suggested a possible new tool in the battle against the epidemic of obesity in the United States, but we now know that the role of leptin is much more complex.

Leptin levels can vary in association with a number of factors independent of adiposity. For example, leptin can differ between populations even after one controls for body fat percentage. The underlying mechanisms for this difference are unclear, although ecological factors may play a role. Leptin is, on average, higher among Americans than in hunter-gatherer populations such as the Ache, and body fat percentage accounts for only part of the difference. Ache women, despite having about 33 percent body fat, exhibit leptin levels that are virtually indistinguishable from those of American women suffering from anorexia nervosa with 7 percent body fat (Bribiescas 2005b). Ache men also exhibit very low leptin levels compared with American men. Lean American men, who have levels of adiposity similar to those of Ache men (about 18 percent), exhibit a significant and strong correlation between leptin and fat percentage, suggesting that regulation of leptin secretion is very sensitive to changes in adiposity (Hickey et al. 1996). Ache men do not show this correlation (Bribiescas 2001c; Bribiescas and Hickey in press).

What is clear, independent of population differences, is sexual dimorphism in leptin, even after controlling for women's greater adiposity. Why do we see these sex differences? They are almost certainly due to some form of sexual selection during our evolutionary past. Since reproduction demands much more energy from females than from males, it makes sense that females would have evolved to be more efficient at storing fat. For males, although fat is less important for reproductive function, it may play a central role in warding off infections and increasing survivability during lean times. Indeed, having too much fat can suppress testosterone levels.

Muscle

If fat is a tissue that gets you through the lean times and is crucial to survivorship, skeletal muscle may be seen as investment in reproductive effort in males. Do we see trade-offs between survivorship and reproductive investment in nonhuman organisms? Yes. As noted in the previous chapter, this sort of trade-off has been documented in those ubiquitous laboratory subjects fruit flies and flatworms. In mammals, by contrast, spermatogenesis requires negligible metabolic investment. The minor metabolic cost of spermatogenesis is clearly indicated by the robustness of sperm production in humans even under the most taxing energetic circumstances. Therefore, we would expect minimal selection for the human male reproductive system's sensitivity to energy balance.

An important metabolic trade-off in many vertebrates is between sexually dimorphic muscle mass, a reflection of somatic investment in reproductive effort, and fat storage or increased immune competence. Phenotypic correlations, however, can confound trade-offs between survivorship and reproductive effort, particularly in organisms characterized by high paternal investment in offspring. Under these conditions, fat deposition can be viewed as contributing to both survivorship and reproductive effort if paternal survivorship is directly correlated with offspring survivorship (see, e.g., Fowler et al. 1994).

In many mammalian species, seasonal changes in body mass are common in association with female receptivity. Male squirrel monkeys *(Saimiri oerstedi)* increase their body mass by 20 percent during the breeding season. Most of this weight consists of fat and water around the head, neck, and shoulders. The gain is attributed to increased testosterone and its subsequent conversion into estrogen (Boinski 1987). Male mandrills *(Mandrillus sphinx)* exhibit distinct changes in another sexually dimorphic characteristic, coloration, in association with seasonal testosterone shifts (Setchell and Dixson 2001). More generally, primate males invest more in sexually dimorphic muscle tissue than primate females (Bolter and Zihlman 2003; Zihlman and McFarland 2000).

So why the emphasis on investing in muscle in mammalian (indeed many vertebrate) males? Almost certainly, sexually dimorphic muscle provides males with greater competitive ability and perhaps enhanced attractiveness to potential mates. So if muscle has these advantages, why don't we see a world full of muscle-bound males? As with most benefits, there is a cost. In men, approximately 20 percent of BMR is derived from skeletal muscle tissue (Elia 1992). Brain tissue also accounts for about 20 percent, but unlike skeletal muscle, brain tissue mass cannot fluctuate during periods of energetic stress. Skeletal muscles can atrophy during periods when the body is using more energy than it takes in—an important mechanism of energy regulation with significant adaptive consequences (Henriksson 1992). Not only does muscle catabolism lower the amount of tissue that needs to be maintained, but it liberates amino acids that can be used as energy resources. Stress hormones such as cortisol play a significant role in muscle catabolism during periods of acute energy stress (Crowley and Matt 1996). Testosterone also keenly influences male allocations of somatic investment through its regulation of muscle anabolism and metabolism.

Testosterone

There are very few endogenous materials in the human body that when distilled are classified as controlled substances by the U.S. government. Testosterone is a schedule three controlled substance. Why? Testosterone is a potent anabolic agent that also can have undesirable effects on the immune system and the prostate. Whether or not large doses of testosterone affect brain function is an open question. Fortunately, under normal circumstances the body produces relatively small quantities of the hormone. The hypothalamus seems to do a good job of monitoring testosterone production and shutting it off to keep men from being completely awash in the stuff. (However, for those foolish enough to self-administer exogenous testosterone, all bets are off.) So why does the male body produce testosterone? Most likely to augment some aspects of reproductive effort even if it seems to compromise survivorship.

The effects of testosterone on human somatic tissue have been well documented. For example, the endocrinologist Shalender Bhasin has conducted numerous experiments illustrating the anabolic effects of testosterone on muscle in younger and older men. Moreover, he and his colleagues have demonstrated that testosterone supplementation does not have to be extreme to have its effects (Bhasin et al. 2001b; Herbst and Bhasin 2004). In addition to the anabolic effects, administration of testosterone stimulates fat catabolism and redistribution of adipose tissue (Wang et al. 2004b). Testosterone and other androgens also stimulate muscle anabolism by increasing protein synthesis and glucose uptake.

A noteworthy example of the effects of testosterone on body composition and energy use emerged from an experiment on patients with muscular dystrophy (MD). Clinicians were hoping to find a way of slowing or halting the muscle wastage that is common in this disease. Three months of testosterone administration to men with MD as well as controls resulted in changes in somatic composition including weight gain due to greater lean body mass and less adiposity, suggesting fat catabolism. Moreover, changes in body composition were associated with increases in BMR of 13 percent in the controls and 7 percent in MD patients. The increased BMR in both groups was attributed primarily to the significant increase in lean body mass (10 percent in MD men, 11 percent in controls) in response to the anabolic effects of testosterone (Welle et al. 1992).

Clearly, from the results for the control group, testosterone has a significant ability to regulate energy use in healthy men. Does this suggest that these men would have compromised survivorship? It's unknown. We cannot conduct experiments on the hypermetabolic effects of testosterone on survivorship in humans. However, it will be interesting to see the mortality data on the flood of men taking testosterone supplements today. But we'll leave that topic for a later chapter.

Data from other vertebrates strongly suggest that heightened testosterone leads to increased reproductive effort, changes in somatic composition in the form of fat catabolism, higher metabolic rates, and compromised survivability (Buchanan et al. 2001; Ketterson et al. 1992). For example, testosterone implants in male desert lizards *(Sceloporus jarrovi)* were associated with increased male mortality (Marler and Moore 1988). Males with implants spent significantly more time in courtship display and aggressive territoriality than sham-treated males. Increased predation risk was not the cause of higher mortality; instead, higher metabolic costs were proposed as an important source of mortality in the males with implants.

In total, the implication of these investigations is that human male metabolic investment in spermatogenesis is minimal (see Figure 7). The bulk of energetic investment is therefore available for various aspects of somatic development. Decisions about allocation of energy between somatic tissues such as skeletal muscle and adipose tissue represent investment in survivorship or reproductive effort. The evidence suggests that augmentation of the ability to grow skeletal muscle reflects investment in reproductive effort, although the increased metabolic costs may

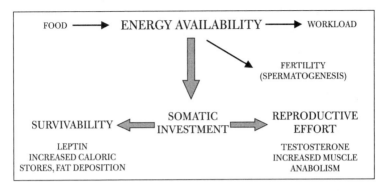

Figure 7 A model of human male reproductive ecology.

compromise survivorship. Investment in relatively inexpensive adipose tissue augments survivorship, perhaps by enhancing the ability to ward off infection (Campbell et al. 2001; Muehlenbein and Bribiescas 2005).

For testosterone to be viewed as an energy-regulating mechanism that has important functions in human male life histories, it must be demonstrated that testosterone varies under differing ecological circumstances. Most of the information we have on testosterone comes from clinical data collected in developed areas of the world. Despite the value of these data, there is ample reason to believe that men living sedentary lives in industrialized societies do not represent the total range or the most common pattern of human variation. Anyone who has traveled to developing parts of the world has probably noted that people living in these regions tend to be shorter than well-fed tourists. If we were to use American or other industrialized-world standards of growth as the most common state of humans, we would be mistaken. Similarly, hormone function varies between as well as within populations.

Some years ago, an undergraduate working on his senior thesis decided to investigate testosterone levels in average men sampled on the street. Armed with a wad of $1 bills, he camped out in an urban square near his campus and proceeded to offer the dollars to men who walked by in return for samples of their saliva. From these saliva samples he could measure testosterone levels. What he found coincided with the results of other researchers (Zitzmann and Nieschlag 2001), that is, salivary testosterone levels varied by about a factor of ten, even after the effects of age were controlled for (Colfax 1990). Subsequent surveys of the range of testosterone levels and profiles in the United States and other industrialized countries have made it clear that some men have consistently and significantly higher testosterone levels than others.

One source of variation is age, something that will be covered in a later chapter. Briefly, testosterone levels do change with age, with physical ramifications such as declines in muscle mass, greater fat deposition, especially around the midsection, possible changes in sexual function and motivation, as well as potential changes in psychological well-being. However even apart from the influence of aging, men exhibit a very wide range of variation in testosterone levels. Why? Clinicians and other scientists are unsure, but there are some factors that merit close scrutiny.

A notable study of male reproductive endocrine variation was conducted by Dan Spratt and Bill Crowley at Massachusetts General Hospital in Boston. In their investigation, they attempted to obtain fine-grained and detailed information on how the HPT axis (the hypothalamus, the pituitary gland, and the testes) functioned, almost minute by minute, in the daily life of a given man. Recruiting twenty healthy men between the ages of eighteen and forty with no record of pubertal or endocrine disorders, Spratt and Crowley arranged to have the subjects continuously connected to blood-drawing tubes. Every few minutes, a valve would be turned and a blood sample would be obtained for analysis of several key hormones. Their findings were quite astonishing. First, not every man exhibited morning highs and evening lows in testosterone, as was common in other investigations. FSH, the hormone that is believed to be central to spermatogenesis, was fairly stable within individuals but varied considerably between subjects. Testosterone exhibited dramatic fluctuations within subjects, sometimes dipping into pathological ranges at night (Spratt et al. 1988).

Testosterone around the World

Testosterone levels are twice as high in American men as in men of many other populations around the world (Bribiescas 1996; Bribiescas 2001a). In addition, the range of variation among Americans is quite extreme, with tenfold differences between the highest and lowest men without any apparent pathological effects.

Most clinical research has focused on the physiology of people living in western industrialized societies, where sedentary lives and virtually unlimited opportunities to eat are the norm. Consequently what has been considered "normal" has been based on the physiology of western individuals. However, progress in sample-storage methods and more sensitive assays have allowed us to make significant strides in our knowledge of the reproductive endocrinology in other populations which may be more representative of most humans in the history of our species.

Hints of population differences in testosterone were first noted some time ago. Surveys of rural black African men reported higher levels of estrogens than those of white African men (Davies 1949). Later investigations reported similar findings, suggesting that indigenous African men had higher levels of estrogens than Europeans, although many of these results have failed to be replicated (Bersohn and Oelofse 1957). What would be considered ethnic differences persist in the literature today, although there is no consensus about the source of such differences (Ellis and Nyborg 1992; Heald et al. 2003).

With the advent of salivary steroid testing, data from remote populations accumulated steadily (Campbell 1994). The robustness of salivary samples, which can be stored without refrigeration, made this methodology ideal for biological anthropologists working in remote locations (Ellison 1988; Lipson and Ellison 1989). The Ituri Project, which investigated the ecology and physiology of central African foragers (Efe) and horticulturalists (Lese) made use of salivary steroid measurements (for a review see Bailey and Aunger 1989). The men exhibited significantly lower salivary testosterone levels than American urban controls (Bentley et al. 1993; Ellison et al. 1989a). In addition, men of other African populations such as !Kung, residents of the Okavango region of Namibia, and those living in Turkana, Kenya, were found to have salivary testosterone levels significantly lower than western controls (Christiansen 1991b, 1991a; Campbell et al. 1995).

Men of other populations also demonstrate significantly lower salivary testosterone levels than Americans, including the Ache of Paraguay (Bribiescas 1996, 1997), Tamang and Kami men of Nepal (Ellison and Panter-Brick 1996), and the Aymara of Bolivia (Beall et al. 1992). And none of these populations has levels so low that we would consider them pathological or subfertile (Galard et al. 1987). They simply seem to be in the lower range of normal testosterone variation.

Testosterone variation between men in westernized, industrialized societies and those in less developed regions of the world is not well understood. Indeed, it is a question that has hardly been addressed. Nonetheless, we can derive several possible reasons for these differences from the clinical literature. There may be issues related to pathology or disease: the general energetic burden of disease as well as the immunological response may impose a significant energetic cost that can potentially suppress testosterone (Muehlenbein 2004; Muehlenbein and Bribiescas 2005). Environmental considerations such as dietary intake, activity, energetic status, and other lifestyle variables also need to be considered. Other factors such as pollution are also potential sources of variation and merit further investigation (Selevan et al. 2000; Sram 1999).

General energetic stress is a possibility. As is readily evident from studies of women, caloric restriction or energetic expenditure can suppress reproductive hormones. However, men's bodily responses to these stresses are not the same as women's. In women, salivary progesterone can decline significantly in response to modest exercise and caloric restriction (Ellison and Lager 1986; Lager and Ellison 1987). Not so in men, except rarely and under the most taxing circumstances (Skarda and Burge 1998). For example, a comparison of sedentary men and long-distance runners revealed no significant differences between the two groups in any reproductive hormones or sperm parameters (Cumming et al. 1986). However, some studies of exercise have shown declines in testosterone (for a review see De Souza and Miller 1997).

Studies of caloric restriction have also yielded ambiguous results. Urinary testosterone in obese men undergoing a ten-day fast exhibited a modest decline but along with it an increase in the production of the gonadotropin FSH (Klibanski et al. 1981). A regimen of severe dietary restriction (600 calories daily) resulted in modest declines in both testosterone and FSH (Hoffer et al. 1986). Short-term fasting affects the production of GnRH and the secretion of LH, FSH, and testosterone (Röjdmark 1987a, 1987b). Outside clinical settings, there have been very few studies of caloric restriction and male reproductive function under natural conditions. A comparison of LH and testosterone among Jordanian men undergoing the Ramadan fast and controls revealed no significant difference in LH but *higher* testosterone in the fasting group (El-Migdadi et al. 2004). But a study of wrestlers showed higher testosterone in those who were not fasting (Booth et al. 1993).

Data on long-term (weeks, months) caloric restriction is scarce. The anthropologist Gillian Bentley and colleagues (1993) reported a decline in salivary testosterone in association with a lean hunger season among a horticultural central African population (the Lese). However, testosterone remained low as food became more available and men gained weight. In total, it can be stated that modest to severe acute energetic stress can affect male reproductive hormones, but that no investigations have reported values that would indicate lower fertility. For women, energetic stress leading to lower ovarian steroid levels can result in subfecundity. The same is not true of men. However, hormonal priming during adolescence due to chronic energetic stress may underlie population differences in testosterone. Among nonindustrialized populations, lower testosterone may be an adaptive response to minimize the potential costs of muscle anabolism (Bribiescas 1996, 2001a.)

As we have seen, men's levels of reproductive hormones such as testosterone vary widely both within and between populations around the world. This may be the result of lifestyle factors that adjust the male reproductive hormone system to accommodate less food and more physical activity. What is clear is that natural selection has surely affected the evolution of testosterone and other hormones in order to balance the needs of reproduction and survivorship. The next few years will undoubtedly witness interesting developments in male hormone research.

9

Men and Medicine

GETTING sick is no fun. If the infection doesn't make you feel terrible, your body's response to the offending agent surely will. Sometimes illness is not due to infection but to some malfunction. On occasion, the very mechanisms that are meant to protect you from infection can lose their ability to distinguish between your own body and any invading microorganism. The monitoring and regulation of immune function are extraordinarily complex and expensive endeavors. Given the malleable and quick-adapting nature of infectious agents, it is logical that the immune system would have been a target of strong selection. The metabolic costs of rallying an immune response or maintaining vigilance against infection are very difficult to quantify, but mounting evidence suggests that immune responses impose a significant metabolic burden on mammals (Raberg et al. 2002). The mode and manner in which males and females invest in immune function differ in important ways, each sex bearing its own particular energetic burdens.

Human males are at a disadvantage compared with human females when it comes to keeping themselves free from parasites and other infectious agents (Beery 2003). Vertebrate males consistently exhibit a compromised ability to fend off infections (Møller et al. 1998; Moore and Wilson 2002). How can something as important as immune function be less efficient in one sex? There are several possibilities. Perhaps males simply get into the infectious line of fire more often. Males' behavior may expose them to greater risk of infection—think of little boys playing in the muck. Second, perhaps specific activities such as hunting in hunter-gatherer groups draw energetic resources away from immune function and make males more vulnerable to infectious agents. Finally, simply being endowed with male physiology may somehow involve a less competent immune system. Sex-specific hormones, for example, may make the male immune system less than optimal. This aspect may involve the effects of androgens or the lack of estrogens.

The associations between the neuroendocrine and immune systems are immensely complex, and the two often overlap (Grossman 1989). A complete overview of our understanding of endocrine/immune function would fill an entire book. But if we view these systems from the perspective of the host (the human body) and the parasite (the infectious agent), patterns emerge that shed light on the physiological and life history trade-offs that evolved in human males.

First, it is important to understand that infectious processes involve competing agendas between the host and the parasite. The infectious agent needs to survive, reproduce, and disperse using resources garnered from the host. However, the agent cannot simply harvest such resources at will. It must deal with the host's defense mechanisms and determine the rate at which resources can be harvested. Harvesting too much at once risks killing the host, which would not be good for the parasite. What is a bug to do? Initially, it was thought that infectious agents tended to evolve toward a harmony with the host in order to obtain a constant supply of resources. The idea was that killing one's host is usually not a good idea. Therefore the level of virulence—the overall effect of an infectious agent on the probability of killing the host—would tend toward nonlethality. Occasionally a rogue agent might emerge that killed its host, but such exceptions would not be very adaptive.

More recently, evolutionary biologists have shown that under conditions that are becoming more common in contemporary societies, pathogen virulence can increase dramatically and ultimately lead to infectious agents that kill their hosts without compromising their own fitness. These conditions include higher population densities, more efficient travel between parts of the globe, expansion into previously unpopulated regions that may harbor previously uncontacted pathogens, and the widespread and less than discriminate use of antibiotics (Ewald 2004). Infectious disease can be best understood from the perspective of an evolutionary biologist knowledgeable about life history theory (Wedekind 1999).

Ornithologists have long known that steroids such as cortisol and testosterone often compromise the ability of male birds to fend off parasites and other infections. In some birds the effect is evident in their plumage. Male cardinals exhibit less vibrant colors when they are heavily infested with parasites (Hamilton and Zuk 1982). Testosterone supplementation exacerbates this effect, so it is no surprise that testosterone levels often decline in proportion to parasite or infection load (Folstad and Karter 1992). The idea is that higher testosterone incurs a greater metabolic burden. When parasite-infested cardinals are given supplementary testosterone, the increased metabolic burden results in less vivid colors. Under natural circumstances, cardinals infested with parasites lower their endogenous testosterone in order to decrease the burden on their metabolisms.

But stresses on immune function are complex and differ according to the pathogenic challenge presented. Pathogens range from bacteria to viruses, zooparasites, and other agents such as fungi. Each pathogen elicits a unique immune response that is differentially sensitive to sex-specific hormonal environments. Perhaps for this reason, the scientific literature includes some contradictory and confusing data (Muehlenbein 2004). Nonetheless, the majority of research does strongly support the contention that sex hormones, specifically testosterone and estradiol, play important roles in immune function (Klein 2000, 2004).

Testosterone Poisoning?

Men seem to get sick from infections more often and more severely than women. And they are more likely to die from them. This may not be completely due to less effective immune responses; it may in part be a function of behavioral factors such as a tendency not to seek medical attention when one is sick (Galdas et al. 2005). The same patterns observed in industrialized countries are also found in less developed parts of the world. For example, among Australian Aboriginal populations, men are more susceptible than women to a number of illnesses such as heart disease (Brown and Blashki 2005). Testosterone is just one of the many male-specific factors that contribute to greater vulnerability to disease. A plethora of investigations of a variety of vertebrates have shown that testosterone can inhibit an individual's ability to ward off infection (Schmid-Hempel 2003).

The effects of testosterone on immune function are readily evident in mammals that are seasonal breeders. In males of these species, testosterone production is characterized by months of quiescence followed by sudden high levels of the hormone. And the males' susceptibility to infection and morbidity is highest during the breeding season.

Male squirrels exhibit lower immune function, higher testosterone, and significant weight loss during the mating season. However, food supplementation prevents immunosuppression and weight loss, supporting the idea of trade-offs between survivorship and reproductive effort (Bachman 2003; Boonstra et al. 2001). Rhesus macaques also exhibit a seasonal pattern of immunosuppression that is probably associated with concurrent changes in testosterone (Mann et al. 2000).

Testosterone supplementation of men exhibiting a wide variety of disorders such as HIV and pulmonary infections does not seem to induce a significant decline in immune function (Bhasin et al. 2000; Casaburi et al. 2004; Kohut et al. 2003). However, these men were well nourished. Perhaps a more valid experiment would be to increase testosterone in undernourished men to avoid the potentially confounding effect of energetic status. Such an experiment would be potentially risky to the subjects since testosterone supplementation might further compromise their immune systems. To date, no such experiments are available. However, catastrophic infections that induce extremely high metabolic demands do indicate a central immunosuppressive role of testosterone (Spratt et al. 1993). Supportive evidence from healthy well-fed men is lacking; perhaps data will emerge in a few years as healthy men now taking testosterone supplements start to age.

The German physician Jörge Schröder and his colleagues (1998) compared mortality rates of men and women who suffered infections (sepsis) following surgery. They found striking contrasts, with a mortality rate of 70 percent for the men but only 26 percent for the women. Modest differences were associated with sex hormones such as testosterone and estradiol. Comparable findings have associated testosterone with greater mortality, although for women some emphasize the role of estrogens over the immunosuppressive effects of androgens (Oberholzer et al. 2000; Offner et al. 1999).

In more natural settings, testosterone has been found to be positively associated with level of parasitemia *(Plasmodium vivax)* in Honduran men presenting themselves for treatment for malarial symptoms. But compared with healthy controls, infected men had lower testosterone and higher cortisol. Perhaps malarial infection compromises a man's ability to produce testosterone. Or perhaps the body decreases testosterone production as part of its response to the infection, in order to lower the hormone's detrimental effects on the rallying of an immune defense (Muehlenbein et al. 2005). In female mice infected with malaria (a different strain from the Honduran men, *Plasmodium chabaudi*), injection of testosterone compromises the liver's ability to mobilize a defense against the disease (Krucken et al. 2005). In male mice, chronic *P. chabaudi* infection lowered testosterone levels but did not seem to affect the ability to father pups (Barthelemy et al. 2004). It may be that because of the relatively low cost of spermatogenesis, immunological energetic stress, including the energetic burden of rallying an immune response, is less likely to affect male fertility than it is to affect other somatic investments such as the building of skeletal muscle tissue (Muehlenbein and Bribiescas 2005).

Fat Chance

Among men who are not obese or overweight, weight loss has been shown to be associated with mortality. In a large-scale study of Japanese-American men, weight loss was positively associated with increased mortality, even after controlling for unhealthy behaviors such as smoking (Iribarren et al. 1995). The allocation of somatic investment in human males may be reflective of life history trade-offs. Adiposity and leptin may reflect investment in survivorship (Bribiescas 2001a, 2001b), which may be traded off against energetic investment in immune function and muscle-building investment that reflects reproductive effort (Muehlenbein 2004).

A growing body of evidence suggests that leptin, the hormonal reflection of adiposity, is intricately involved in the rallying of an immune response to infection (La Cava and Matarese 2004; Lord et al. 1998). For example, induced pneumonia infections caused significantly higher mortality in leptin-deficient mice than in normal wild-type mice (Mancuso et al. 2002). Sex differences in leptin are ubiquitous across populations despite significant variation in absolute levels (Bribiescas 2001c). It is therefore likely that male leptin physiology is simply not as robust as female in supporting immunological defenses. Greater adiposity leads to higher estrogen levels because fat cells convert testosterone into estradiol. The differences in leptin physiology between men and women may mirror the differences seen between leptin-deficient and wild-type mice.

Why is there a male immunological deficit? Perhaps it is a trade-off involving higher testosterone or lower estradiol. Induced increases in men's estradiol levels do not increase leptin, but testosterone and leptin are often inversely related, most likely as the result of the catabolic effect of testosterone on fat cells (Luukkaa et al. 2000, 1998). Human males reflect the common pattern of mammalian immune function in that morbidity and survivorship tend to take a back seat to metabolic investment in reproductive effort.

Evolutionary Medicine

Those who treat illness have a different perspective on it from those who attempt to understand the evolution of pathogens and the physical responses of the afflicted (Nesse and Williams 1994; Williams 1991). The contrasts between those who treat disease and those who wish to understand it from an evolutionary perspective involve the basic concepts of proximate and ultimate causation. Physicians, in general, are trained to deal with the immediate causes of illness, the proximate or nuts-and-bolts causes. If something in a patient's body is not functioning as it should, a doctor will troubleshoot the ailment and prescribe a course of treatment. This is understandable since no patient wants to sit in the emergency room feeling miserable and listening to a lecture on the evolution of influenza.

While this perspective is appropriate for immediate treatment, in the long run, a more complete understanding of how influenza became a threat may prevent future infection. Evolutionary biologists and anthropologists examine concepts of illness as well as the selective factors that have resulted in the evolution of disease. They focus not only on the physiology of disease but also on public health issues related to infectious and degenerative illness. Physical responses to infection involve traits that may have adaptive significance. One of the best-known examples is sickle-cell anemia. Evolutionary biologists have shown that the evolutionary roots of sickle-cell anemia lie in demographic changes involving sedentary living, agriculture, and protection against malaria. These changes have resulted in selection for the recessive alleles that cause anemia in homozygous individuals, those who inherit the sickle-cell allele from both parents. Heterozygous persons, those with only one copy of the allele, are not anemic, but their red blood cells are altered in a way that protects them against the protozoan that causes malaria (Wiesenfeld 1967).

Therefore, the question "What causes sickle-cell anemia?" would yield different but equally correct answers from a physician and an evolutionary biologist. The physician would state that sickle-cell anemia is caused by the malformation of red blood cells, which results in clotting and inefficient oxygen transport. The biologist would attribute the origins of the disease to changes in human settlement patterns that resulted in higher population densities and more standing water that promotes the propagation of the anopheles mosquito, which passes the malaria-causing protozoan to humans. Heterozygous individuals with one sickle-cell allele have some measure of protection against malaria and are more likely to survive to reproductive age, have offspring, and pass this allele to the next generation. Both those without the allele and those with two doses of it are at a disadvantage. People without the allele are more vulnerable to malaria, and homozygous individuals with two sickle-cell alleles are selected against since they tend to die from anemia before they can reproduce. The allele is maintained in the population by the heterozygous individuals who have the protection against malaria while avoiding the anemia.

Evolutionary biologists also recognize that many illnesses and conditions are the result of antagonistic pleiotropy, the existence of traits that are detrimental to survivorship later in life but are maintained within a population because of their positive effects on reproduction. Many of the health conditions that are of concern to contemporary male humans are likely to have emerged from pleiotropic effects.

A variety of health concerns are unique to men. Some probably have a long evolutionary history, while others are relative newcomers that arose from demographic changes. Among the health issues of interest to men in industrialized societies are prostate cancer, the potential development of a male oral contraceptive, and androgen supplementation therapy.

The Male Cancer

The prostate is a strange structure. About the size of a walnut, it is involved in several aspects of male reproductive and sexual function. Located just behind the groin region, it surrounds and engulfs the urethral tract. The primary functions of the prostate are to produce prostatic fluid, the medium to which sperm is added before ejaculation, and to aid in the creation and maintenance of erections. The prostate also is the site of the most common form of cancer unique to men. The causes of prostate cancer involve many factors, including genetic heritability, environmental considerations such as diet and activity, and, perhaps, carcinogenic substances in the environment.

One main characteristic of the prostate gland is the proliferation and maintenance of numerous binding sites for androgens such as testosterone and DHT. If you remember, DHT is a more potent androgen that binds and activates testosterone receptors more efficiently than testosterone does. Five alpha reductase, the enzyme responsible for converting testosterone to DHT, is abundant within prostatic tissue. Consequently, the growth and maintenance of prostate cells rely extensively on androgen stimulation. Indeed, upon first diagnosis of benign prostatic hyperplasia (enlargement of the prostate) or prostate cancer, one of the first courses of treatment is to introduce androgen suppressors in order to block the hormones' stimulatory effects on prostate cells (Anderson 2003).

Prostate cancer affects men in a nonrandom fashion: certain groups are much more susceptible than others. For example, prostate cancer seems to occur more frequently in men living in industrialized regions than in those living in less developed areas (Quinn and Babb 2002). For urban-dwelling men in industrialized societies, prostate cancer is one of the leading causes of mortality, despite the fact that in some industrialized parts of the world, such as Japan, this cancer is much less prevalent than elsewhere (Kurihara et al. 1989). In the United States, prostate cancer afflicts a greater proportion of African-American men than would be expected by sheer chance (Powell 1997). In fact, it afflicts African-American men at a rate up to twice that of white men (Ross et al. 1992; Reddy et al. 2003), and is often more aggressive than in white men (Berger et al. 2006). In addition, young U.S. black men have been reported to exhibit significantly higher testosterone levels than their white counterparts (Ross et al. 1986). We know that androgens such as testosterone and DHT play a central role in fostering prostate cancer growth (Isaacs 1996), but the relationship between lifetime exposure to testosterone and risk of prostate cancer remains unclear (Gann et al. 1996; Eaton et al. 1999).

The genetic contribution to susceptibility to prostate cancer is also unclear. There is evidence that familial heritability is associated with risk of developing prostate cancer (Shibata and Whittemore 1997), but the pathway through which these genes operate is unknown (Page et al. 1997; Spaas and Bagshaw 1990). Potential genetic factors may involve hormonal receptors within the prostate, the way prostate cells grow, or patterns of programmed cell death (apoptosis) (Zhang et al. 2002). Studies of the impact of genetic variants that influence prostate development and androgen receptors are few, and they have produced limited evidence that such genetic variants are a causal source of variation between populations. A comparative investigation of polymorphisms in androgen receptors between two populations with divergent prostate cancer risks (China, low; Australia, high) revealed no significant differences between these populations, suggesting that environmental or lifestyle considerations play a more prominent role than such polymorphisms (Jin et al. 2000). Environmental factors such as diet, specifically fat intake, and activity patterns warrant close attention given the impact they have on the neuroendocrine system (Key 1995; Schröder 1996; Whittemore et al. 1995).

What underlies the high rate of prostate cancer among African-Americans? Is it genetic? Perhaps partly. However, Nigerian men exhibit less aggressive bouts of prostate cancer as well as lower testosterone levels than African-American men living in Washington, D.C., implying that genetic variants may have less effect than lifestyle differences (Odedina et al. 2006). In addition, an investigation of whether hereditary prostate cancer risk is higher for African-American men than for their white counterparts failed to uncover any significant differences between these groups (Cunningham et al. 2003). An intriguing recent investigation did report a common genetic variant that was associated with prostate cancer and was almost twice as common among African-American populations than among Europeans (Amundadottir et al. 2006). What is clear is that other steroid-specific cancers such as breast cancer also exhibit population differences and that in looking for their causes environmental and lifestyle differences should be considered (Jasienska and Thune 2001; Jasienska et al. 2000). There is some evidence that developmental and environmental conditions early in life may be particularly important for breast cancer, while those later in life may be more important for prostate cancer (Shimizu et al. 1991).

Why does prostate cancer even exist? At first glance, one might hypothesize that any traits that lead to prostate cancer would be selected against because of their obvious detrimental effects on men's health. Moreover, prostate cancer often leads to sexual dysfunction, as the result of the disease itself or of the treatment (Fitzpatrick et al. 1998), making it less likely that men with prostate cancer will reproduce and thus making it even more likely that the traits that foster the cancer should be selected out of the population. However, this is where the feature that the evolutionary biologist George C. Williams dubbed "antagonistic pleiotropy" comes into play. Selection does not necessarily favor survivorship if issues related to reproductive success are involved. Traits that favor reproduction can be maintained in a population even if they undermine survivorship later in life. In addition, traits that are detrimental to reproduction can be maintained in a population if those traits are not expressed until later in life, after most reproduction has occurred.

The hormones related to prostate disease, testosterone and DHT, are central to male reproduction. Without these hormones, men would not be able to manifest the secondary sexual characteristics that make them viable sexual partners. Testosterone and DHT also play secondary roles in spermatogenesis. While sperm can be produced in the absence of testosterone or DHT, the process is most efficient when they are present. So these androgens are beneficial early in life in promoting reproduction. In later years, however, they impose a burden on survivorship by encouraging excessive cell growth in the prostate. It is important to note, though, that prostate cancer seems to be relatively rare in nonwestern populations, and that whether it had any significant impact on human male evolution is unknown. Nonetheless, the relationship between prostate cancer and testosterone does highlight the importance of recognizing the effects of antagonistic pleiotropy when attempting to understand the origins of a disease.

Testosterone Supplementation

Besides the scandals on the sports pages concerning illegal use of anabolic steroids, legal, medically prescribed hormone treatment for men has become a popular topic in the general media. Testosterone and other androgen supplementation regimens are now commonly available to older men wishing to feel more libidinous, to enhance their sense of well-being, or to improve their somatic condition by having more muscle and less fat (Bhasin and Buckwalter 2001). As men age, libido wanes, muscle withers, and fat tends to accumulate even in those with the healthiest diets and the most rigorous exercise regimens. Some of these changes result from declines in testosterone with age (Gray et al. 1991; Harman et al. 2001). Comparison of the rates at which testosterone is cleared by the body in older and younger men indicates that older testes are less capable of producing testosterone (Wang et al. 2004a). Consequently, it has been suggested that supplementation of testosterone or other androgens may be therapeutic. Under what conditions is testosterone supplementation appropriate? Are men really suffering from testosterone deficiencies? In hypogonadism and in cases where trauma or disease necessitates removing the testes, it is reasonable to consider testosterone-replacement therapy to offset the loss of endogenous hormone production. But what about men who are simply growing older?

And apart from the motivation for seeking treatment, how much testosterone is necessary or optimal? The effects of testosterone on various physiological functions such as muscle building, libido, and cognitive function do not necessarily share the same dose-dependent responses. The effects of variations in receptor density, sensitivity, responsiveness, and rates of saturation are not well understood. Shalender Bhasin and his research team attempted to address these issues by controlling for endogenous hormone levels, administering various preselected dosages of testosterone, and observing the effects on somatic composition, strength, cognitive function, aspects of libido, and other physiological functions. Caloric and nutrient intake were controlled and monitored, as were activity levels and even behaviors that might influence testosterone levels such as competitive interactions. The results indicated that somatic composition, that is, the balance of fat-free mass and fat mass, changed in a linear dose-dependent manner: muscle mass increased while fat mass decreased as dosages of testosterone rose. Strength was equally responsive to higher hormone dosages. However, there were no differences in libido or cognitive abilities between men receiving different dosages.

These results are in accordance with previous studies that observed no changes in libido in association with testosterone administration above the minimum for normal testicular function (Buena et al. 1993). Even when very high doses were administered to normal men, levels of sexual activity did not increase, although some parameters of libido exhibited a modest rise (Anderson et al. 1992). The one group in which testosterone has been found to have dose-dependent effects on libido is made up of men with very low testosterone levels due to a developmental disorder, disease, or trauma. In these hypogonadal men, administration of increasing dosages of testosterone resulted in dose-dependent responses in frequency of erections, ejaculation, and arousal. However, once testosterone levels reached the lower range found in men with normal gonadal function, the dose response diminished (Davidson et al. 1979; O'Carroll et al. 1985; Salmimies et al. 1982).

Intramuscular injection is the most common mode of administration of testosterone, although dermal patches, gels, and even nasal sprays have also been developed (Kuhnert et al. 2005; Mazer et al. 2005; Ko et al. 1998). Patches in particular have received attention as a noninvasive and discreet method that allows a gradual release of the hormone, whereas injections result in a spike in testosterone levels followed by gradual decline. Also, the use of patches, unlike injections, does not require special patient training or administration by medical professionals. Administration of testosterone using scrotal patches has shown promise of being a relatively safe and effective manner of increasing testosterone levels in appropriate patients (Behre et al. 1999).

However, is testosterone supplementation a good idea? What are the implications of giving testosterone to aging men, possibly increasing their risk of prostate cancer? Is there such a thing as an optimal testosterone profile? As indicated earlier, there is a broad range of variation in testosterone levels within and between populations. There are few studies of the effects of testosterone supplementation in populations characterized by differing mean hormone levels. One study compared the effects of testosterone enanthate injections in Chinese and non-Chinese men. HDL cholesterol declined in non-Chinese men but remained unchanged in Chinese men. Liver transaminases rose in Chinese men, an indication of liver damage or abnormality, but were unaltered in non-Chinese subjects (Wu et al. 1996).

The biological anthropologist Gillian Bentley (1994), who investigated population variation in response to oral contraceptive administration in women, found that different female populations exhibited a wide range of hormone clearance rates, resulting in members of some populations receiving a greater overall dosage of the hormone than others. She noted that physicians and health workers need to increase their awareness that equal dosages to different populations may result in unexpected differences in circulating hormone levels. Testosterone infusion experiments did not reveal differences in clearance rates between white and Asian men (Wang et al. 2004a), but more research is clearly needed.

The Male Pill

Today, as in the past, women bear most of the responsibility for preventing unwanted pregnancy. With the large leaps in our knowledge of male reproductive endocrine physiology, researchers have sought to develop an oral contraceptive for men, one that inhibits spermatogenesis but does not interfere with other hormonally driven functions such as libido and muscle tone. This research has met with some success, and clinical trials have been conducted in various populations (Anderson et al. 2002; Brady et al. 2006; Zhang et al. 2006). The successful development and implementation of a male contraceptive will depend on several factors, some physiological, some related to attitudes, behavior, and public health policy.

First the physiological issues. To what degree does spermatogenesis need to be inhibited for a contraceptive to be considered acceptably effective? Female oral contraceptives are about 98 percent effective when used properly, so let's use that benchmark. How do we alter or control spermatogenesis to achieve a 2 percent probability of conception, assuming no contraceptive use by the female partner? There are several possible options.

Hormone blockers exist that halt production of GnRH and FSH or block their receptors. Without FSH or functioning FSH receptors, sperm production falls off quite rapidly. But is this effect enough to thwart conception? Not entirely. In women, mutations that block FSH function cause complete inability to conceive. But men with similar mutations are still capable of producing some sperm and retain a small but significant possibility of fathering children (Kumar et al. 1997; Layman and McDonough 2000).

In men, clinicians are challenged by the multifaceted role of testosterone and gonadotropins. Simply shutting down GnRH production, while probably effective in inhibiting spermatogenesis, also results in declines in testosterone and its secondary effects on muscle maintenance and libido. The primary strategy is to inhibit spermatogenesis and replace any lost testosterone with exogenous supplementation. Initial trials of a proposed male contraceptive involved injections of testosterone enanthate in order to induce a negative feedback response by the hypothalamus, thereby shutting down production of GnRH and subsequently, LH and FSH while maintaining secondary sexual characteristics such as muscle mass. However, it became evident that men's sensitivity to these injections varied significantly. In some men spermatogenesis was immediately arrested, while in others it merely declined by a limited amount. Also, some men exhibited adverse responses such as muscle wasting and unwanted changes in sex drive (Wu et al. 1996).

It is quite likely that fine-tuning the dosages and agents needed to block spermatogenesis while maintaining testosterone function is well within the capability of the clinical research community. So why hasn't a male oral contraceptive appeared on the market? It's probable that the greatest obstacle to the development and deployment of a male pill lies in getting men to use it. Even if a safe and effective hormonal male contraceptive were produced, one that completely negated the possibility of conception while avoiding unwanted side effects, it is unclear whether it would achieve widespread adoption (Waites 2003; Weston et al. 2002).

In the World Health Organization's programs to promote male contraceptives, clinicians and public health officials have had some success with men in urban, industrialized societies such as the United States, particularly with men whose female partners were unwilling or unable to use contraception because of contraindications such as high blood pressure. In nonindustrial populations, however, several concerns became evident. For example, men in Scotland, South Africa, and China expressed hesitation about using oral contraceptives because of fear of unwanted effects on libido or other characteristics considered masculine (Martin et al. 2000). Clearly, any successful promotion of male oral contraceptives would have to be based on an understanding of the social factors involved, and awareness of and sensitivity toward cultural diversity would be key elements in its success.

So how do male behavior and male evolutionary physiology fit into the development of a pill for men? While women have strong evolutionary and physiological reasons to wish to regulate family size and avoid unwanted pregnancies, what would motivate men to use oral contraceptives, given that they do not bear the physical costs of reproduction? One motivation would probably arise from awareness that men would be held accountable for investing in their children. That is, men in countries that require fathers to pay child support might be more interested in oral contraceptives than men who do not expect to be held accountable. Men who are committed to investing heavily in nurturing their progeny might also be interested in oral contraceptives as a way to space out the births of their children. Finally, men considering vasectomies might opt for an oral contraceptive instead as a way to leave open the possibility of having children in the future.

In any case, it is unlikely that men's use of oral contraceptives would ever supersede women's. The metabolic costs of childbearing—as well as the social and economic burdens of unwanted pregnancy—fall largely on women. Therefore, the incentives to use oral contraceptives are much weaker and narrower for men than for women. Nonetheless, the development of a male contraceptive pill might alleviate some of the burden on women, especially women who are unable to use oral contraceptives themselves because of high blood pressure, headaches, or other health concerns.

The Old Guard

TIME IS A FINITE RESOURCE. Unlike energy, time can't be stored for later use. It can only be allocated to one pursuit or another. In addition, everyone and everything gets older. But getting older is not the same as senescence. Aging is commonly associated with the simple passing of time, while senescence is the rate and manner in which an organism declines in function over time. Some organisms senesce more rapidly than humans, some more slowly. In sexually reproducing species such as humans, males and females tend to senesce at different rates and in different ways. Even within a species or an individual, various physiological processes decay in dissimilar ways. For example, reproductive function and liver function may senesce at different rates. Senescence is a life history trait that is subject to natural selection and has undergone significant evolutionary change in all species, including humans.

Why Get Old?

Charles Darwin left us long ago, having lived to the ripe old age of seventy-three. Not bad for a human male during the nineteenth century.

Today you can visit his mortal remains at Westminster Abbey and toast his life at one of the many fine pubs in London. You can no longer invite Darwin to join you for a pint, but if you happen to be in Australia it is still possible to meet someone who may have known Darwin, or at least met him. As of 2005, Harriet was 175 years old. Harriet, a female Galapagos tortoise, is said to be one of three members of her species that were collected by Darwin himself during his famous trip to the Galapagos Islands on the *HMS Beagle*. The trio were initially named Tom, Dick, and Harry, but Darwin turned out to be wrong about Harry's sex. Harriet lives quietly (how else?) at the Australia Zoo. Tom and Dick, like Darwin, are no longer among us.

Amazingly, Harriet is not the oldest tortoise we know of. Two others lived longer but have since died. We commonly think of tortoises and two-thousand-year-old bristlecone pine trees as the epitomes of longevity, but in practical terms some organisms never age. The evolution of senescence as a trait that is subject to natural selection requires organisms that can be distinguished into discrete age categories. Those that cannot be categorized by age seem to live forever. Which organisms are blessed with such characteristics? Unicellular organisms such as bacteria cannot be placed into discrete age categories since each individual is a clone of its originating cell.

Age classification requires the separation of somatic and germ cell lines (Rose 1991). Organisms that have separate germ and somatic cell lines are exposed to the processes of senescence, more specifically, to time-associated changes in somatic cells that ultimately lead to death. In all practicality, germline cells are immortal since they are passed from generation to generation through the production of gametes. For bacteria, the cell is both the somatic cell and the source of reproductive duplicates. For the rest of us, senescence is a fact of life.

The evolutionary biologist Michael Rose (1991) has shown through elegant laboratory based selection experiments in drosophila (fruit flies) that the manner and rate at which an organism senesces can be changed through evolutionary processes. In other words, the traits that affect the lifespan of an organism, after controlling for accidents and other extrinsic factors, are subject to natural selection and have evolved through Darwinian processes. Acting as the selective force, Rose selectively bred individuals who lived longer than control flies, thus showing that components involved with longevity and different rates of senescence are genetically encoded. This does not mean that there is a single "longevity gene," but it does imply that various genetic traits contribute to the rate at which an organism degrades. Rose's results however do not mean that one can produce a fruit fly that can live as long as Harriet. As with all other traits that are subject to selection, phylogenetic constraints limit the range of variation that can be expressed. Nonetheless, senescence is a life history trait that is central in determining how much time an organism has to conduct its business of growing and reproducing.

If senescence is a function of natural selection, then why haven't humans evolved the ability to live two, three, or four hundred years? Obviously, some constraint has held us back. There are basically four theories that attempt to explain the evolution of senescence. The first theory invokes an explanation that is related to the idea of group selection, a topic we discussed in Chapter 1. The idea is that senescence evolved as a way to avoid overtaxing an organism's environment and burdening future generations—more simply, to avoid overpopulation. But this explanation has two basic flaws. First, we know that arguments based on group selection are often evolutionarily unstable. That is, if a mutation came along that made individuals selfish, those mutants would propagate at the expense of the more altruistic individuals, and the genes for altruism would disappear from the population. Second, there is little evidence that senescence is an important source of mortality in the wild. There are some exceptions among semelparous species (organisms that have only one major reproductive event in their lifetime) such as Pacific salmon, in which death closely follows reproduction. But for the most part, animals in the wild do not die of old age.

A second theory, proposed by George C. Williams in the 1950s, is that some organisms simply burn themselves out. That is, species with fast metabolic rates, such as small mammals, are burdened with a cost of having fast metabolisms (Williams 1960). This is known as the "rate of living" theory. As in a car with a fast idle, the engine—or in the case of a mouse the heart, liver, and so on—breaks down sooner because of greater wear and tear. Perhaps this is why mice live for only a couple of years while elephants and humans live for decades. And, in fact, limiting rodents' caloric intake slows their metabolic rates, decreases cellular damage caused by oxygen use and metabolism, and results in longer-lived rodents (Couzin 2004). But does the same apply to other animals such as primates? Initial results from a longitudinal investigation indicated that decreasing the caloric intake of rhesus monkeys by 30 percent significantly lowered mortality and morbidity (Mattison et al. 2003). Such findings concur with known detrimental effects of fast metabolisms (Stadtman 2001). Not surprisingly, this life lengthening effect of calorie limitation has received much attention in the clinical and popular press as well as received generous grants for further research.

But there are issues that remain to be resolved. For example, the rule that smaller animals live shorter lives has some notable exceptions. Important exceptions include bats, opossums, naked mole rats, and even some birds such as parrots. These species tend to live much longer than one would predict from their body size. What accounts for their longevity? Lower externally caused mortality may have something to do with it. For example, bats and naked mole rats have very few predators because of their environments. Bats occupy the air (unusual for a mammal), and naked mole rats live in remote, isolated underground colonies in southern African deserts. The lack of predation on these species may allow them to delay reproduction and thus to devote more of their energetic resources to survivorship.

Another question raised by the rate-of-living theory is this: If restricting caloric intake extends the life of an organism so significantly, why hasn't there been stronger selection for physiological mechanisms that promote dietary restraint? Most likely the answer is that caloric restriction also causes inhibited reproductive function, especially in females, which is not a very adaptive trait. A weakness of the research is that animals in laboratory experiments on caloric restriction were protected from immunological challenges, not a very natural or realistic setting. There is plenty of evidence that energetic stress, such as that caused by caloric restriction, dampens immunocompetence and suppresses survivability. Thus when considering the effects of organisms' metabolic rates one must consider other processes as well.

The third theory suggests that as part of the normal everyday processes of life, genetic defects accumulate as the result of cell replication. Additional damage is caused by factors such as cosmic rays and any variety of toxic substances found in the environment. To deal with these challenges, organisms are endowed with repair mechanisms, but eventually the accumulation of deleterious defects wins out because extrinsic sources of mortality limit the strength of selection for the repair of genetic damage. It takes energy to repair DNA, so if environmental hazards dictate that an individual probably has a 99 percent chance of being picked off by a predator within, say, five years, there won't be much selection to optimize DNA to last beyond those five years. Even organisms living in a sheltered environment, such as a laboratory, won't live much longer than expected because of the phylogenetic inertia imposed by ancestral populations that spent their lives dodging environmental hazards. Since animals tend to die well short of their maximal life span, deleterious mutations that accumulate over time are subject to weak or no selection. Ultimately these bad mutations accumulate and you die, even if you have the good luck to avoid lions and meteors.

What keeps organisms from evolving super-efficient mechanisms of gene repair? Perhaps the repair of genes, tissues, and organs is subject to the same energy constraints as other processes. The fourth theory, which was proposed by the gerontologist Tom Kirkwood and came to be known as the "disposable soma theory," suggested that the necessity of allocating energy to processes such as growth and reproduction limited the resources that could be used to keep up with accumulating genetic and somatic defects (Kirkwood 1977, 1999). This idea dovetails nicely with life history theory. It seems quite likely that differences in reproductive investment between species could indeed contribute to the differing patterns of senescence we observe.

How long does it take to evolve such differences? It does not take many generations of differential selection pressure due to extrinsic mortality to see changes in senescence patterns in subsequent generations. Michael Rose demonstrated this in fruit flies. But drosophila can produce a tremendous number of generations over a short period of time, say several months, which is the reason they are such a favorite among genetic researchers. What about mammals? The biologist Steven Austad hypothesized that selection related to senescence resulting from differences in predation risk could be evident in a mammalian species within a relatively short period of time, on the order of hundreds of years. His idea was that, within a species, populations that were exposed to more environmental threats needed to grow and reproduce at a faster rate than populations with fewer threats. Because of the life history trade-offs imposed by the limited availability of energy, organisms that grew and reproduced faster would incur greater costs. Such costs should compromise maintenance, resulting in lower survivorship and more rapid senescence.

Austad hypothesized that communities of the same species living in the same general environment but with different risks from predators would exhibit different rates of senescence. He quantified senescence by measuring physiological factors such as changes in collagen, a fibrous tissue. His experiment involved examining two populations of opossums belonging to the same species. One population lived on an island without predators, the other on the mainland, where there were predators such as coyotes. He found that opossums on the island reproduced more slowly and exhibited fewer physical signs of age-related degeneration. In addition, the island opossums were in generally better health than their mainland counterparts that had to contend with predators. Without predators, island opossums had the luxury of producing fewer offspring per unit of time and investing more heavily in each offspring as well as in their own survivorship (Austad 1993).

Do we see similar patterns in humans? Perhaps. Rudi Westendorp and Tom Kirkwood (1998) reported intriguing findings gleaned from historical data on the English aristocracy. They discovered that age at first birth was lowest in women who died early and highest in women who lived longest. Moreover, when they considered only women who had lived to be over sixty years of age, they found a negative relationship between longevity and number of children but a positive association between longevity and age at first birth. In other words, having children early and often meant you were likely to have a shorter life.

Investment in aspects of life that take energy away from somatic maintenance shortens life spans. In the case of opossums, more energy per unit of time is devoted to reproduction as a result of greater environmental hazards. Mainland opossums need to reproduce more quickly and invest more energy in reproduction at any given time since they have a lower chance of surviving to the next day than their island counterparts. This may account for the anomalous life spans of bats and some marsupials. These organisms enjoy relatively low risk of environmental danger. Therefore, they can invest energy in self-maintenance, reproduce slowly, and live longer than the norm for animals of their size.

Recall from the previous chapter Williams's concept of antagonistic pleiotropy: genetic traits that have a reproductive benefit early in life but dire consequences as one ages would still be selected for since older individuals tend to be past their reproductive period and therefore not subject to strong selection forces. Modern research on evolutionary gerontology taps into all these theoretical perspectives and nicely illustrates the importance of life history trade-offs in the biology of aging.

When one considers the cumulative roles of reproductive effort, extrinsic mortality, and metabolic rate, some clarity begins to emerge about the evolution of human male senescence. High levels of extrinsic mortality lead to, and are products of, greater reproductive effort. Why such an investment in reproductive effort? Again, for males, the potential fitness payoffs of behavior and biology that are detrimental to survivorship yet beneficial to reproduction are high. For the most part, limitations on male fertility are imposed by limitations on the males' access to mates. This seems to have resulted in selection for a male physiology that burns fuel at a high rate in order to be prepared to take advantage of opportunities to mate. Whether this is true in contemporary human societies is unclear. What is clear is that these selective forces were active in our immediate ancestral past. As reflected by basic male physiology, higher metabolic rates due to investment in sexually dimorphic tissue surely contribute to rates of senescence we see in most mammalian males, including humans. It is important to recognize that although, as a species, humans have unique patterns of senescence, they also seem to share many features of senescence with other mammals and, indeed, other vertebrates.

Human Senescence

It is very likely that reproductive effort, extrinsic mortality, and sex-specific challenges to energy allocation strategies have played major roles in the evolution of human senescence. Humans produce many more offspring than other mammals of our size such as gorillas and orangutans. Moreover, these offspring require high levels of care over a long period of time, much longer than is usual in mammals of our size. If we extrapolate from Austad's findings on opossums to humans, it would seem that during the course of human evolution, environmental sources of mortality such as predation became less pronounced, allowing humans to grow longer and create many high-quality offspring that need significant parental care.

It has long been suggested that male provisioning of females, changes in diet, foraging efficiency, and the evolution of large brains are all related to the evolution of longer lives (Aiello and Wheeler 1995; Lovejoy 1981). These factors in turn may have permitted females to increase their reproductive output. More efficient acquisition of food resources may have augmented ovarian function and fertility. Male provisioning of food for women and children, probably in the form of meat acquired through hunting, significantly curtailed women's predation risk while allowing them to devote more time and energy to offspring care. (For an alternative view that emphasizes the role of tubers over meat, see Laden and Wrangham 2005; Wrangham et al. 1999.) This is not to say that women came to live lives of leisure. Women in foraging societies work incredibly hard and contribute significant amounts of food to their households. However, over millennia, small differences in availability and expenditure of energy can lead to important changes in fertility and mortality (Hill et al. 2001; Kaplan et al. 2000).

Men and women exhibit distinct differences in the manner in which their physiology responds to the passage of time. For example, on average, women live significantly longer than men. Even when male high-risk behaviors are factored out and the dangers of childbirth are taken into account, men still die sooner. Cross-cultural studies have shown earlier male mortality to be a widespread phenomenon, suggesting that the underlying cause of faster male senescence is physiological or perhaps genetic and physiological (Buettner 1995; Hill and Hurtado 1996; Tomasson 1984; Waldron 1985). There are many factors that may contribute to sex differences in mortality, but among them, testosterone is central. Surgically or chemically castrated men live longer than intact men (Owens 2002). There are probably many physiological factors behind these differences, including the effects of testosterone on immune function and the overall metabolic costs of maintaining more skeletal muscle mass than women do (Moore and Wilson 2002; Muehlenbein 2004). But testosterone is probably not the sole factor in these differences. Estrogen, an immunological bolstering agent, may provide some protective effects against infection (Verthelyi 2001).

Human males exhibit specific signs of senescence. The most prominent are loss of muscle mass, increased adiposity, and other changes such as a decline of cognitive abilities and a general deterioration of health. In addition, the hypothalamus, the area of the brain that regulates many hormonal tasks, may become less capable of regulating energy allocation at older ages. A decline in the efficiency of energy allocation would imply that even under favorable circumstances, such as the availability of more than sufficient food, the body might not be able to apportion resources appropriately to its own somatic functions.

The Aging Male

Senior Moments

Cognitive function declines with age, although there is significant variation in the rate and severity of mental deficits associated with senescence. Men in general are more susceptible to brain damage than women, although there is some debate in the literature (Braun et al. 2001). There seem to be only minor differences between older men and older women in regard to cognitive function (Barrett-Connor and Kritz-Silverstein 1999). However, following a stroke, men are slower to recover than women. Estrogens and progesterone, which have some cognitive-enhancing effects, may play a role in these differences. Indeed, progesterone and estradiol, a major estrogen, have been proposed as therapeutic agents to improve and speed up recovery from stroke and other brain damage in men (Stein 2001).

The relationship between estradiol and cognitive function in women has received a significant amount of support. However, the role of testosterone and estradiol in male cognitive health is less clear. Endogenous testosterone levels are not associated with variation in cognitive function in older men (Yaffe et al. 2002). Moreover, administration of exogenous testosterone to older men has been found to raise testosterone levels but to yield only modest improvements to cognitive function (Kenny et al. 2002; Wolf and Kirschbaum 2002). Mood and depression, however, may be related to testosterone. Testosterone supplementation of hypogonadal and older men sometimes alleviates depressive symptoms, but results have been mixed. Part of the problem is that the relationship between testosterone and depression is often confounded by other factors such as stress, smoking, and alcohol use (Carnahan and Perry 2004).

The brain undergoes actual physical changes as humans age. We are all familiar with the plight of those with Alzheimer's disease, but not all changes in the brain are so severe. Overall, there are few differences between men and women regarding age-related changes in brain volume and neural mass. Older men and women exhibit similar declines, relative to body size, in gray and white matter (Resnick et al. 2000). But shifts in brain organization with age involve differential changes in the size and shape of specific brain structures. E. V. Sullivan and colleagues (2002) performed magnetic resonance imaging (MRI) on the brains of 215 elderly men twice, four years apart. They found a significant thinning of the corpus callosum, as well as larger increases in the size of ventricles, fluid-filled spaces within the brain. Increases in ventricular space imply a loss of brain matter. Although more data are needed regarding sex differences in age-related changes in cognition and brain structure, the present picture indicates that men are at greater risk than women of changes in brain function as they get older.

BULKING DOWN

As men age, muscle becomes more difficult to develop and maintain while fat becomes harder to burn. Since men tend to have more skeletal muscle tissue than women, decreases in muscle-building efficiency and increases in ease of atrophy are of particular relevance to men. What causes these changes and why? Do men become more sedentary and eat more as they get older? Well, probably. But getting fatter and less muscular is about more than just lifestyle changes; it is also related to distinct changes in hormones that regulate somatic composition.

Large epidemiological studies of men in two American cities, Boston and Baltimore, have found that testosterone tends to decline by about 1 percent per year after the age of forty (Gray et al. 1991; Harman et al. 2001). Whether these declines are universal across populations is an open question (Ellison et al. 2002). But in aging American men, waning testosterone coincides with decreases in muscle mass and increases in adiposity. Declines in testosterone have several effects. First, declines reduce muscle building. With less muscle, the basal metabolic rate decreases, contributing to the preservation and deposition of more fat cells. More fat cells mean more conversion of testosterone into estradiol. Increases in estradiol result in more fat deposition and greater negative feedback to the hypothalamus, which can further suppress production of LH and testosterone (Hayes et al. 2000). The sum of all these effects is love handles and potbellies. So what happens when we replace an aging man's declining testosterone? Supplementation over a period of three months has been found to promote greater muscle mass and less adiposity (Tenover 1992). Indeed, Bhasin and colleagues (2005) have reported that older men are as responsive as younger men to the muscle-building effects of supplemental testosterone, implying that age-related changes in the number or the sensitivity of testosterone receptors may be minimal.

However, as noted earlier, clinical data from western industrialized societies may not reflect the full range of variation around the world. While testosterone decreases with age in American men, populations in other parts of the world, such as the Ache of Paraguay, exhibit a much less steep decline or none at all with age (Bribiescas 1996; Ellison et al. 2002). But perhaps male aging still manifests itself through somatic deterioration. The anthropologists Robert Walker and Kim Hill (2003) assessed physical performance and aging among Ache men. They conducted a series of physical performance tests including pull-ups, sprints, grip strength, and push-ups. They also assessed VO2max, a measure of the body's maximum capacity to utilize oxygen. They found, not surprisingly, that physical performance peaked in men's early twenties and tended to decline slightly but steadily thereafter. These are the same changes in strength that are commonly observed in industrial populations.

But, as Walker and Hill point out, strength is not the only requirement for male-specific activities that are vital to forager livelihoods. Skill is crucial as well. Among foraging populations, hunting is central to the lives of males of all ages. I recall watching an Ache toddler, no older than three, harassing a small lizard with a toy bow and arrow made for him by his father. The interest in hunting is instilled early and maintained throughout a male's life. While many hunter-gatherer groups, including the Ache, do less foraging today than they once did because of deforestation and cultural assimilation, it is still possible to glean insights from their lives into behaviors, such as hunting, that were probably influential factors in the evolution of the classic human traits of large brains, dependent offspring, and highly fertile females. While physical strength may be a useful male trait, aging may provide men with skills that can only come with experience, skills that result in real-world advantages.

The anthropologist Irven DeVore used to have a coffee mug emblazoned with the phrase "Age and treachery will always triumph over youth and vigor." Although I have no empirical evidence of the political ruthlessness of older men, skills acquired with age may become more important than physical strength as men grow older. The findings of Robert Walker and colleagues (2002) about the effects of age, skill, and physical prowess on hunting success among Ache men illustrate this point. Early assumptions by anthropologists were characterized by notions of strapping young men dominating the hunt and reaping reproductive benefits. But is hunting success due to being stronger or more skillful? Obviously all Ache males need some semblance of physical health to go out into the Paraguayan forest, find prey animals, and bring them down. But the amount of food Ache hunters bring home peaks when the men are in their thirties, a decade after their peak of physical strength. Walker and his colleagues note that while men in their early twenties are physically stronger and run faster, the older men consistently exhibit greater success at finding prey and return with more meat. Years of experience in the forest are more important in determining hunting success than simple brute strength.

It is interesting to speculate that male hunting ability, increased brain size, and decreased sexual dimorphism are interrelated and all perhaps emerged at the time of *Homo erectus,* a species characterized by these three traits. It would appear that, in our near ancestors, males took advantage of the large brain and evolved the ability to bring down game through guile and wits. And it may be that the growing importance of hunting contributed to the evolution of a large brain. Why? Perhaps the selective advantage was increased survivorship. Neanderthals seem to have suffered many severe head and neck injuries, perhaps because their hunting strategy was to get up close and personal with large, dangerous game (Berger and Trinkaus 1995). While some minimal strength is necessary, it is more important to live long enough to gain experience.

MALE MENOPAUSE?

We all know that women undergo a distinct period of reproductive senescence. Menstrual cycles lengthen, become irregular, and eventually
cease altogether, signifying the end of a woman's ability to conceive and
bear children. What makes menopause unusual in the animal kingdom
is not the process of reproductive senescence itself but the length of
time that women live after the reproductive stage of their lives. Approximately one-third of the human female life span is postmenopausal. This
is in contrast to most other organisms, in which reproductive and overall senescence are highly correlated; that is, once a female has lost her
ability to reproduce, death usually soon follows. Some have argued that
forms of menopause are evident among other primates such as baboons, chimpanzees, and rhesus macaques, as well as in other mammals
(Cohen 2004; Gould et al. 1981; Packer et al. 1998). However, postreproductive life in these organisms is quite brief compared with that of
human females.

Over the past several years, attention has been drawn to the possible
existence of a male menopause. In order to determine the validity of
claiming the existence of a male menopause, we must identify the defining characteristics of female menopause. First and foremost, we understand that reproductive potential declines with the depletion of viable ova. All primordial oocytes, the cells that eventually become ova,
are produced during fetal development, before a girl is born. At that
stage they undergo one division and then stop in a sort of biochemical
deep freeze. Starting at a young woman's first menstruation, cohorts of
oocytes are stimulated into growth and possible ovulation.

Men's fertility is not constrained by a predetermined number of gametes. The production of sperm, once it begins at puberty, is a continuous process that does not undergo an abrupt cessation with age. The process of spermatogenesis does become somewhat compromised with age, as is evident from the increase in genetic abnormalities in the sperm of older men (Plas et al. 2000), but the supply of gametes is not exhausted. Barring severe illness or disorder, men retain the ability to produce viable sperm well into their later years.

But being capable of producing viable sperm does not necessarily mean an older man will retain the ability to inseminate a woman. The ability to produce and maintain an erection can be significantly affected by age—an aspect of senescence that explains much of the enormous success of sildenafil citrate (better known as Viagra) and similar drugs (Kaiser 1999). Age-related declines in testosterone and increases in body fat, which can further suppress testosterone, contribute to the incidence of erectile dysfunction. However, age alone is a significant influence as well, with the risk of erectile dysfunction increasing about 8 percent per year in men between the ages of forty-five and sixty (Kratzik et al. 2005). Besides factors such as diabetes and smoking, the most salient aspect of senescence that is related to erectile dysfunction is degeneration of the vascular system. This should not be a surprise since erections rely on vascular function to shunt blood flow to the penis. Increases in vascular plaque and thickening of the vascular walls cause higher penile blood pressure, which can inhibit erections (Caretta et al. 2006).

Research on erectile dysfunction outside the United States is sparse but growing (Nicolosi et al. 2003). Even among the most remote populations, risk of being unable to get or sustain an erection is associated with age. Among Ariaal pastoralists of northern Kenya, Peter Gray and Benjamin Campbell (2005) found that the incidence of erectile dysfunction was significantly higher in men over the age of sixty than in men fifty and younger.

As noted in an earlier chapter, erections occur when signals involving nitric oxide reach the smooth muscles around the corpora cavernosa of the penis. These muscles relax, allowing blood to flow into the spongy tissue of the corpora cavernosa, which expands as it fills with blood. Drugs such as sildenafil citrate work by mimicking the effects of nitric oxide and stimulating the smooth muscles around the corpora cavernosa, increasing blood circulation to the penis (Argiolas and Melis 2003).

There are some modest challenges to treating erectile dysfunction with sildenafil (Viagra). Although older men taking sildenafil face no apparent increased risk of heart disease, their risk of a drop in blood pressure does increase. Men who are taking nitrates for heart disease are not advised to take sildenafil because it can cause a sudden decrease in blood pressure. In general, though, older men tend to tolerate sildenafil quite well with few side effects (headaches, flushing, stomach discomfort). Nevertheless, some mild risk is associated with successful treatment of erectile dysfunction, since older men are at greater risk of heart failure during and after sexual activity (Salonia et al. 2005).

Other aspects of male reproductive function do change with little or no effect on fertility. Testosterone levels tend to drop over a man's lifetime because of declines in LH, as well as increases in SHBG, a protein that binds with testosterone and other sex hormones in the bloodstream. Increases in FSH, a hormone responsible for stimulating spermatogenesis, are also evident (Gray et al. 1991; Harman et al. 2001). But keep in mind that age-related changes in testosterone do not seem to be universal (Ellison et al. 2002). And none of these factors suggests that men undergo an abrupt cessation of reproductive potential analogous to women's menopause.

While American men exhibit a decline in salivary testosterone with age, Ache men do not. Congolese and Nepali men fall between these two extremes (Ellison et al. 2002). What this suggests is that male reproductive senescence, as indicated by declines in testosterone, is a phenomenon of western industrialized societies. As noted earlier, populations that work hard and have chronic energetic deficits tend to have lower testosterone levels. In addition, a study of elderly urban Japanese men found that salivary testosterone tended to level off after the age of forty and remained quite stable well into the tenth decade of life—at least for men who were in relatively good health; less healthy men did not participate in the study (Uchida et al. 2002). This finding lends credence to the idea that overall health and somatic condition are as important in maintaining testosterone levels as is age itself.

But despite retaining the ability—or at least the sperm count—to father offspring, most men do not have children after the age of fifty, around the same time as women's menopause. A study of contemporary German men found that male fertility, or the number of children fathered annually, peaks between the ages of thirty and thirty-five (Plas et al. 2000). Data from Ache men, too, show that peak male fertility is during the mid- to late thirties, with an abrupt decline after this period even though there is no evidence of compromised reproductive physiology. Besides the influence of the age of female partners, the reasons for this decline in male fertility may be sociological and physical. Ache women may prefer younger men, older Ache men may lose the physical capacity to attract and keep a mate, or more likely a combination of both.

So why don't more men father children later in life? As we have seen, sperm production in older men may be lower and their sperm may have more genetic abnormalities. A recent French investigation reported that a man's age was consistently associated with a greater risk of failure to conceive regardless of the woman's age. Risk of failure to conceive was especially high in men over the age of forty (de La Rochebrochard et al. 2006). But impaired sperm quality is not a full explanation. Older men's access to sexual partners may be reduced if women choose younger mates. In many cultures, a man's access to resources such as cash or cattle may help him attract sexual partners. In polygynous cultures, wealthy older men who can afford more wives may have greater reproductive success (Cronk 1991b). However, even if an older man has access to sexual partners, his ability to impregnate them is likely to be compromised, given the prevalence of age-related erectile dysfunction. Therefore, fatherhood at advanced ages remains relatively unusual.

MALE LONGEVITY AND MENOPAUSE

But why do women live so long after they've had their last menstrual period? One well-known idea is the "grandmother hypothesis" formulated by the anthropologist Kristen Hawkes. Childbearing is an inherently risky endeavor, often resulting in maternal health problems or death. But the obvious necessity of reproducing supersedes these risks, and women have children. Hawkes suggests that the risks associated with childbearing increase with age because of women's declines in overall health and robustness. Therefore, it may be adaptive for women to stop having children before the risks become too great and invest their energy in other individuals who carry their genes, such as grandchildren. Several tests of this hypothesis have been conducted with mixed results. Hawkes reported that among Hadza women of Tanzania grandmothers provided a significant amount of supplementary food to daughters and grandchildren (Hawkes et al. 1997). In addition, child-care provided by grandmothers may improve infant survivorship and therefore increase the grandmothers' inclusive fitness. Similar patterns were found among maternal grandmothers and children in rural Gambia (Sear et al. 2000).

But showing that grandmothers have a positive effect on the health and well-being of grandchildren does not provide sufficient evidence that menopause is an adaptive trait. The clinching finding would be evidence that cessation of reproduction and investment in grandchildren would yield a greater fitness payoff than if the woman continued to produce her own offspring. This is a tall order since the number of grandchildren produced or surviving as a direct result of grandmothers' intervention would have to be high enough to overcome the discounting of genetic distance. Direct offspring, say daughters or sons, share 50 percent of the mother's genes, while grandchildren from daughters only share 25 percent (and grandchildren from sons may not be related at all because of paternal uncertainty). That is, for menopause and grandmotherly investment to be more advantageous than continued childbearing, for every offspring not produced by a woman her actions as a grandmother would have to result in the survival of three extra grandchildren. But to date there is no evidence that grandmotherly contributions offset the grandmothers' fitness losses from ceasing their own reproduction at a relatively early age.

To reject the grandmother hypothesis at this point would be premature. Although compelling supportive evidence has not been forthcoming, further testing of Hawkes's idea is called for. Indeed, the grandmother hypothesis is one of the few testable ideas that have given researchers a useful approach to the question of menopause. Another idea is the "mother hypothesis": that women may cease reproduction in order to increase the chances that their last-born children will survive and reproduce (Peccei 2001). Another possibility is that the depletion of a woman's oocytes is a basic physiological constraint that is not adaptive. Accumulation of genetic defects in oocytes is associated with a woman's age (Kline and Levin 1992). So it may be that menopause is simply the exhaustion of the oocyte supply: that the ability to store oocytes is not indefinite, with biochemical constraints limiting the duration of viable storage. But this is not a very satisfying explanation since it fails to address the issue of women's postmenopausal longevity.

The anthropologist Frank Marlowe has proposed a novel approach that, in the context of male life histories, merits discussion. Marlowe argues that the decoupling of physical strength from crucial abilities that are central to acquiring mates and increasing offspring survivorship led to the evolution for longevity in males. Selection for male longevity must also involve selection of relevant genes. Since these genes are shared with offspring during fertilization, both male and female children will inherit these longevity genes from their fathers. In essence, the idea is that females are included with the selection for greater longevity in males, with the consequence that females outlive their own reproductive systems (Marlowe 2000). In the context of this "patriarch hypothesis," the grandmother hypothesis may take on a new perspective. If female longevity increases because of selection for longevity in males, it may make sense for older women to devote resources to their grandchildren in an effort to increase their inclusive fitness since they themselves cannot reproduce anymore. They may simply be making the best of a less than perfect situation. However, it is equally valid to postulate that a long postreproductive period evolved to ensure that a woman could raise her last-born child to maturity (the mother hypothesis), and that males were the ones who were dragged along with this evolutionary scenario, especially since men suffer higher mortality than women.

WHAT ABOUT GRANDFATHERS?

One widely held image of grandfathers is as benevolent elders, sitting around the village fire or in their rocking chairs and passing on knowledge and wisdom that may be useful to future generations. And yet, as we saw earlier, since men don't go through menopause, many older men are physiologically able to continue producing children of their own rather than investing in their grandchildren. Could there be a "grandfather hypothesis," similar to the other hypotheses just discussed, to explain why many grandfathers do invest in their offspring's offspring? To test such a hypothesis we would have to determine whether grandfathers' contributions to childcare or other family needs enhance the survival of offspring or grandchildren, or, more specifically, whether the fitness payoff to a man of investment in his grandchildren outweighs the benefits he would gain from investing more effort in attracting fertile mates, fathering more children, and helping those children survive. This would be a much more daunting task than testing the grandmother hypothesis since uncertainty about paternity would always fog any interpretations. Few studies have looked directly at grandfathers' effect on their descendants, but there is some compelling evidence suggesting that having a grandfather around may not always be a good idea from a grandchild's perspective.

The anthropologist Cheryl Sorenson Jamison and colleagues (2002) set out to test the grandmother hypothesis by examining demographic records from a period spanning two hundred years (1671–1871) in a rural Japanese community. Their goal was to determine whether the presence or absence of grandparents was associated with differences in childhood mortality. They found a beneficial effect of maternal grandmothers: a child was 35 percent more likely to survive in a household that included its mother's mother. However, they also found that the presence of either grandfather in a household had a negative impact on children's survival. In rural Gambia, Rebecca Sear and colleagues (2000) found no significant effects, positive or negative, of the presence of live-in grandfathers on children's health and condition.

The association between grandfathers' presence and grandchildren's mortality discovered in the Japan study was unexpected, and causal factors are not known. Indeed, the researchers note that their sample sizes are not as large as they would desire. In addition, because of the manner in which infants were counted by census takers in rural Japan, those who died in their first year of life were not included. This is noteworthy given that that first year is a crucial period in which child mortality is at its highest. Nonetheless, the negative associations between grandfathers and children's survival may be indicative of a distribution of family resources that is detrimental to children. The observed effects may not be due to any actions, subtle or overt, on the part of grandfathers.

Much remains to be learned about the evolution of senescence, and about evolutionary medicine in general (Nesse at al. 2006). Physicians are instructed in the aspects of physiology and anatomy necessary for treating illness. Yet medical schools do not teach evolutionary theory. Physicians are not provided with the intellectual tools to formulate useful questions about the development and etiology of disease or the human body's often perplexing reactions to pathogens. Moreover, increased awareness of the full range of nonpathological variation in human biology is needed. If physicians were cognizant of the broad range of human differences in, for example, age-related changes in testosterone, new insights could be gained into male-specific diseases such as prostate cancer. Aside from the medical benefits, evolutionary biology is a ripe area of inquiry for those who wish to understand the evolution of our species and our place in the natural world.

CONCLUSION

The Solitary Male

THE TOPIC OF HUMAN MALE EVOLUTION-
ary biology involves and is interrelated with many
issues that are at the core of science: women's rights, psychology, violence, warfare, socioeconomic conditions, and so on. Does biology explain all of men's behavior? Can biology predict which specific individual is destined to be the next Hitler or the next abusive spouse? On the
basis of the review of male biology presented in this book, the answer is
a resounding no. However, understanding the basic constraints and
challenges of life history that shaped male biology does provide clues
into why conflicts between nations are so prevalent. Men run the world.
There has never been a female American president and only a few other
nations have elevated a woman to their highest office. The political,
economic, and social status of women still needs major improvement.
Are women oppressed, or do men have an advantage when it comes to
obtaining and keeping power? Yes.

Throughout recorded history men have wielded most of the political
and economic power. They have also been at the centers of major conflict, upheaval, and genocide—medieval Europe, the dynasties of China,

Mesoamerican civilizations, it doesn't matter. Conflict in hunter-gatherer societies, when it does occur, also seems to be a male affair. Is this at least partially due to the fact men have generally been the ones to record history? Absolutely. But historical distortion cannot completely explain these consistent patterns. Indeed, male-biased history further illustrates that men tend to hold the reins of power. So where does biology fit in?

Men, in many ways, have been selected to be selfish. Before I get a flood of correspondence skewering me for "justifying" men's selfishness, I would like to state definitively that all men (all people for that matter) bear responsibility for their actions. "Evolution made me do it" is not a valid defense. But consider the different contributions of fathers and mothers to the production of human children. Men don't have monthly menstrual bleeding, they don't get pregnant, they don't give birth, and they don't breastfeed babies. All these activities make demands on women's energetic resources, and pregnancy and lactation involve direct transfers of the mother's somatic resources to the child. Men, in contrast, are not biologically obligated to share their bodily resources with offspring. Indeed, until and unless male humans evolve some way of converting somatic resources into a useful and transferable form of energy, such direct sharing won't even be possible. Some male insects, such as butterflies, do this, but their reproductive biology is very different from that of mammals.

Many men do invest heavily in family and children, of course, and the extended period of care needed by human young gives caring fathers long-term incentive to use their somatic means to provide resources with which to enhance their offspring's well-being. There are wonderful fathers in the world, mine among them. He was nurturing, thoughtful, and a great provider, and he deserves much of the credit for any good traits I may have. (Mom's a gem, too.) But unlike mothers, fathers are not *required by their biology* to provide child support. Every calorie ingested by a human male is his to keep—and to invest, if he sees fit, in pursuits other than protecting and provisioning the younger generation.

Keep in mind that, thanks to the way the human reproductive system has evolved, men can't be positive that they fathered particular children. Being a nurturing father requires making a conscious choice to invest considerable time and energy in children a man *believes* to be his own. In many ways it takes a leap of faith. This is not to say that women, who *can* be certain of which offspring are their own (at least they could during most of the history of our species, before the days of in vitro fertilization and surrogate pregnancy), are completely and always required to invest in children. Under some circumstances abandoning a child, turning it over to others to raise, or even killing it might be adaptive (Hrdy 1999; Hausfater and Hrdy 1984). Not right, not wrong—adaptive. If abandonment and abuse of children can arise in women, whose genetic relatedness to the children they consider their own is assured, it is not a stretch to imagine that such practices would be more common in men.

What does this have to do with wars and other nasty behavior? Species in which fathers invest heavily in caring for offspring, such as penguins, are not also universal altruists. Males (and females) fight, bite, and kill. What may have occurred in the evolution of human males is the merging of three generalized mammalian male traits—the inability to share bodily resources with offspring, contribute to their gestation, or assuredly identify their relatedness—with selection for investment in mating effort, large brains, and sociality. Put these all together and you get a combination that is highly likely to spawn organized warfare. Sociality is a common trait in primates. Large brains are unique to humans, and their evolution almost certainly involved access to high-quality and consistent sources of food as well as lower extrinsic mortality that allowed our ancestors to live long enough to grow and fill such a big brain.

Part of males' investment in mating effort went into building larger bodies and muscularity, qualities that, once developed as a result of sexual selection, were not likely to go unused. Moreover, these traits would not have evolved in a vacuum. Neuroendocrine function would have evolved in tandem for more efficient use of resources, metabolically and behaviorally. Add a large brain, and efficiency would increase even more. That is, males would come to rely on brain power more than muscle power to vanquish their enemies. But females also have large brains. Why aren't they also at the forefront of war or organized violence? What separates males from females in this respect may well be parental assuredness and somatic investment in offspring. The questions facing males are: Are my children really mine? And what do I do with calories and time since I don't have to share them with offspring?

I have always been impressed that Albert Einstein and others were able to arrive at deep and important insights without Apple Powerbooks or any of the high-tech tools used today. They relied heavily on thought experiments, exercises in logic that enabled them to foreshadow results that would be confirmed as the technology to test their ideas was developed. I humbly present the following thought experiment that illustrates the central role of paternity uncertainty in human male evolution. Imagine a scenario in which a man has infinite resources, infinite access to women with whom to mate, and unbridled attractiveness. He also is not constrained by extrinsic mortality. Under these conditions, the potential fitness payoffs for our hypothetical super-male would be enormous. He might father thousands of offspring, or none. And yet, despite all his mating effort and his likelihood of fathering at least some offspring, he would not know for certain which were his or even whether he had fathered any at all. (Yes, over the centuries men have devised many mate-guarding systems to keep rivals away from their women, but our thought experiment allows for the possibility of human "sneaky copulators.") Repeat this experiment over and over and you will still have the same result. The possibility that a man might be unrelated to anyone in the next generation despite such a huge mating effort is a staggering thought. The bottom line is that women who bear their own children (not surrogates) are the only people who can assuredly identify those who are genetically related to them. Men can never have the same assurance. From an evolutionary perspective, males are quite alone.

Without the inhibitions that result from not wanting to harm their offspring or their kin, men might well be less restrained than women in their actions, at a personal and global level. Add the fact that, because men do not need to invest metabolically in offspring, they are able to invest more time and energy in accumulating social, economic, and political power, and one can imagine how male behavioral propensities may have evolved. Men wage war because they can and because the potential costs of such action to their evolutionary fitness may be zero. This thought experiment does not imply that men make such choices consciously. But conscious thought is not necessary for selection to influence men's behavior by shaping their biological nature.

Is this biological nature static, or has it changed since our split from our closest evolutionary relatives? Many of these male propensities are not unique to humans. In their book on the evolution of aggression and violence, Richard Wrangham and Dale Peterson (1996) propose that the evolutionary roots of aggression and violence within men may have originated in the common ancestor of humans and chimpanzees. Male chimpanzees have many similarities with men, including organized violence that is often lethal. However, the evolutionary split between the two species occurred between five and seven million years ago (Patterson et al. 2006). That's a long time. Have humans and chimpanzees evolved any differences in their violent ways? Perhaps. Although rates of lethal violence do not differ between wild chimpanzees and humans, humans have much lower rates of nonlethal violence (Wrangham et al. 2006). There may be hope for us guys after all.

In a classic article on anthropological research into human life histories, Kim Hill (1993) noted that many public policies are based on the erroneous assumption that humans have evolved to maximize longevity—that is, the assumption that our behavior is geared toward allowing us to live as long as possible. While it is certainly true that survival and well-being are important motivational forces, data from human populations such as adolescent males strongly suggest that other motivational factors are at work and that our behavioral physiology is not always inclined toward choices that benefit our well-being. As noted earlier in this book, evolutionary and life history theory propose that all organisms, including humans, are subject to selection pressures that favor optimal lifetime reproductive success—the passing of one's genetic makeup to future generations—not longevity.

When we bear these points in mind, undesirable behaviors associated with poverty and high mortality, such as gang warfare, begin to take on some measure of clarity. If individuals believe, either consciously or unconsciously from a lifetime of observations and cues from their environment, that they have a significant risk of dying young, they may well be likely to engage in high-risk or antisocial behaviors. A healthy respect for legal or moral consequences of one's actions requires the belief that one will have a future. As noted by Father Gregory Boyle, a Jesuit priest who spent many years living and working with the gangs of Los Angeles, a future is exactly what many gang members believe will elude them. After saying mass at the funeral of one of many young men killed by gang violence, Father Boyle stopped to comfort a boy around eight years old. In response to his consoling words, the boy simply shrugged and said, "We're all going to die sometime"—a sobering perspective for a child who hadn't even finished his first decade of life (Fremon 1995).

Future research into human evolutionary physiology will continue to involve the melding of theoretical developments with clinical and field studies. Although most of our understanding of human physiology comes from clinical studies of urban dwellers in the United States and Europe, it is becoming clear that human physiological variation ranges far beyond what has long been accepted in clinical circles. Moreover, understanding human physiology requires quantitative research in areas that have traditionally been overlooked. Evolutionary anthropologists have much to offer in this discussion today and will continue to do so in the future.

REFERENCES

ILLUSTRATION CREDITS

ACKNOWLEDGMENTS

INDEX

REFERENCES

Abbas, S. M., and A. H. Basalamah. 1986. "Effects of Ramadhan fast on male fertility." *Archives of Andrology* 16: 161–166.

Aboitz, F., A. B. Scheibel, R. S. Fisher, and E. Zaidel. 1992. "Fiber composition of the human corpus callosum." *Brain Research* 598: 143–153.

Abzhanov, A., S. Holtzman, and T. C. Kaufman. 2001. "The Drosophila proboscis is specified by two Hox genes, proboscipedia and sex combs reduced, via repression of leg and antennal appendage genes." *Development* 128: 2803–14.

Achenbach, G. G., and C. T. Snowdon. 2002. "Costs of caregiving: Weight loss in captive adult male cotton-top tamarins *(Saguinus oedipus)* following the birth of infants." *International Journal of Primatology* 23: 179–189.

Achiron, R., S. Lipitz, and A. Achiron. 2001. "Sex-related differences in the development of the human fetal corpus callosum: In utero ultrasonographic study." *Prenatal Diagnosis* 21: 116–120.

Adams, M. S., and J. D. Niswander. 1973. "Birth weight of North American Indians: A correction and amplification." *Human Biology* 45: 351–357.

Ahmed, S. F., A. Cheng, L. Dovey, J. R. Hawkins, H. Martin, J. Rowland, N. Shimura, A. D. Tait, and I. A. Hughes. 2000. "Phenotypic features, androgen receptor binding, and mutational analysis in 278 clinical cases reported as androgen insensitivity syndrome." *Journal of Clinical Endocrinology and Metabolism* 85: 658–665.

Aiello, L. C., and C. Key. 2002. "Energetic consequences of being a *Homo erectus* female." *American Journal of Human Biology* 14: 551–565.

Aiello, L. C., and P. Wheeler. 1995. "The expensive tissue hypothesis: The brain and the digestive system in human evolution." *Current Anthropology* 36: 199–221.

Allan, B. B., R. Brant, J. E. Seidel, and J. F. Jarrell. 1997. "Declining sex ratios in Canada." *Canadian Medical Association Journal* 156: 37–41.

Amos, W., S. Twiss, P. P. Pomeroy, and S. S. Anderson. 1993. "Male mating success and paternity in the grey seal, Halichoerus grypus: A study using DNA fingerprinting." *Proceedings of the Royal Society of London, Series B* 252: 199–207.

Amundadottir, L. T., P. Sulem, J. Gudmundsson, A. Helgason, A. Baker, B. A. Agnarsson, A. Sigurdsson, K. R. Benediktsdottir, J. B. Cazier, J. Sainz, M. Jakobsdottir, J. Kostic, D. N. Magnusdottir, S. Ghosh, K. Agnarsson, B. Birgisdottir, L. Le Roux, A. Olafsdottir, T. Blondal, M. Andresdottir, O. S. Gretarsdottir, J. T. Bergthorsson, D. Gudbjartsson, A. Gylfason, G. Thorleifsson, A. Manolescu, K. Kristjansson, G. Geirsson, H. Isaksson, J. Douglas, J. E. Johansson, K. Balter, F. Wiklund, J. E. Montie, X. Yu, B. K. Suarez, C. Ober, K. A. Cooney, H. Gronberg, W. J. Catalona, G. V. Einarsson, R. B. Barkardottir, J. R. Gulcher, A. Kong, U. Thorsteinsdottir, and K. Stefansson. 2006. "A common variant associated with prostate cancer in European and African populations." *Nature Genetics* advance online publication, accessed at *www.nature.com*.

Anderson, J. 2003. "The role of antiandrogen monotherapy in the treatment of prostate cancer." *BJU International* 91: 455–461.

Anderson, K. G., and H. Kaplan. 1999. "Paternal care by genetic fathers and stepfathers I: Reports from Albuquerque men." *Evolution and Human Behavior* 20: 405–431.

Anderson, R. A., J. Bancroft, and F. C. Wu. 1992. "The effects of exogenous testosterone on sexuality and mood of normal men." *Journal of Clinical Endocrinology and Metabolism* 75: 1503–07.

Anderson, R. A., Z. M. Van Der Spuy, O. A. Dada, S. K. Tregoning, P. M. Zinn, O. A. Adeniji, T. A. Fakoya, K. B. Smith, and D. T. Baird. 2002. "Investigation of hormonal male contraception in African men: Suppression of spermatogenesis by oral desogestrel with depot testosterone." *Human Reproduction* 17: 2869–77.

Andersson, A. M., J. Toppari, A. M. Haavisto, J. H. Petersen, T. Simell, O. Simell, and N. E. Skakkebaek. 1998. "Longitudinal reproductive hormone profiles in infants: Peak of inhibin B levels in infant boys exceeds levels in adult men." *Journal of Clinical Endocrinology and Metabolism* 83: 675–681.

Andersson, K. E., and G. Wagner. 1995. "Physiology of penile erection." *Physiological Reviews* 75: 191–236.

Andolz, P., M. A. Bielsa, and J. Vila. 1999. "Evolution of semen quality in north-eastern Spain: A study in 22,759 infertile men over a 36-year period." *Human Reproduction* 14: 731–735.

Apicella, C. L., and F. W. Marlowe. 2004. "Perceived mate fidelity and paternal resemblance predict men's investment in children." *Evolution and Human Behavior* 25: 371–378.

Argiolas, A., and M. R. Melis. 2003. "The neurophysiology of the sexual cycle." *Journal of Endocrinological Investigation* 26: 20–22.

Austad, S. N. 1993. "Retarded senescence in an insular population of Virginia opossums *(Didelphis virginiana)." Journal of Zoology* 229: 695–708.

——— 1994. "Menopause: An evolutionary perspective." *Experimental Gerontology* 29: 255–263.

——— 1997a. "Comparative aging and life histories in mammals." *Experimental Gerontology* 32: 23–38.

——— 1997b. *Why We Age: What Science Is Discovering about the Body's Journey through Life.* New York: Wiley.

Aversa, A., F. Mazzilli, T. Rossi, M. Delfino, A. M. Isidori, and A. Fabbri. 2000. "Effects of sildenafil (Viagra) administration on seminal parameters and post-ejaculatory refractory time in normal males." *Human Reproduction* 15: 131–134.

Bachman, G. C. 2003. "Food supplements modulate changes in leucocyte numbers in breeding male ground squirrels." *Journal of Experimental Biology* 206: 2373–80.

Bagatell, C. J., and W. J. Bremner. 1990. "Sperm counts and reproductive hormones in male marathoners and lean controls." *Fertility and Sterility* 53: 688–692.

Bagatell, C. J., J. R. Heiman, J. E. Rivier, and W. J. Bremner. 1994. "Effects of endogenous testosterone and estradiol on sexual behavior in normal young men." *Journal of Clinical Endocrinology and Metabolism* 78: 711–716.

Bahadur, G., K. L. Ling, and M. Katz. 1996. "Statistical modelling reveals demography and time are the main contributing factors in global sperm count changes between 1938 and 1996." *Human Reproduction* 11: 2635–39.

Bailey, R. C., and R. Aunger Jr. 1989. "Significance of the social relationships of Efe Pygmy men in the Ituri forest, Zaire." *American Journal of Physical Anthropology* 78: 495–507.

Bakker, J., C. De Mees, Q. Douhard, J. Balthazart, P. Gabant, J. Szpirer, and C. Szpirer. 2006. "Alpha-fetoprotein protects the developing female mouse brain from masculinization and defeminization by estrogens." *Nature Neuroscience* 9: 220–226.

Balthazart, J. 1991. "Testosterone metabolism in the avian hypothalamus." *Journal of Steroid Biochemistry and Molecular Biology* 40: 557–570.

Barker, D. J. P. 1998. *Mothers, Babies, and Health in Later Life.* New York: Churchill Livingstone.

Barrett-Connor, E., and D. Kritz-Silverstein. 1999. "Gender differences in cognitive function with age: The Rancho Bernardo study." *Journal of the American Geriatrics Society* 47: 159–164.

Barthelemy, M., C. Gabrion, and G. Petit. 2004. "Reduction in testosterone concentration and its effect on the reproductive output of chronic malaria-infected male mice." *Parasitology Research* 93: 475–481.

Beall, C. M., C. M. Worthman, J. Stallings, K. P. Strohl, G. M. Brittenham, and M. Barragan. 1992. "Salivary testosterone concentration of Aymara men native to 3600m." *Annals of Human Biology* 19: 67–78.

Beck-Peccoz, P., V. Padmanabhan, A. M. Baggiani, D. Cortelazzi, M. Buscaglia, G. Medri, A. M. Marconi, G. Pardi, and I. Z. Beitins. 1991. "Maturation of hypothalamic-pituitary-gonadal function in normal human fetuses: Circulating levels of gonadotropins, their common alpha-subunit and free testosterone, and discrepancy between immunological and biological activities of circulating follicle-stimulating hormone." *Journal of Clinical Endocrinology and Metabolism* 73: 525–532.

Becker, S., and K. Berhane. 1997. "A meta-analysis of 61 sperm count studies revisited." *Fertility and Sterility* 67: 1103–08.

Beery, T. A. 2003. "Sex differences in infection and sepsis." *Critical Care Nursing Clinics of North America* 15: 55–62.

Behre, H. M., S. von Eckardstein, S. Kliesch, and E. Nieschlag. 1999. "Long-term substitution therapy of hypogonadal men with transscrotal testosterone over 7–10 years." *Clinical Endocrinology* (Oxford) 50: 629–635.

Beise, J., and E. Voland. 2002. "Effect of producing sons on maternal longevity in premodern populations." *Science* 298: 317.

Bell, G. 1978. "The evolution of anisogamy." *Journal of Theoretical Biology* 73: 247–270.

Bell, G., and V. Koufopanou. 1986. "The cost of reproduction." In *Oxford Surveys in Evolutionary Biology,* ed. R. Dawkins and R. Ridley, 83–131. Oxford: Oxford University Press.

Bentley, G. R. 1994. "Ranging hormones: Do hormonal contraceptives ignore human biological variation and evolution?" *Annals of the New York Academy of Sciences* 709: 201–203.

Bentley, G. R., A. M. Harrigan, B. Campbell, and P. T. Ellison. 1993. "Seasonal effects on salivary testosterone levels among Lese males of the Ituri Forest, Zaire." *American Journal of Human Biology* 5: 711–717.

Berard, J. D., P. Nurnberg, J. T. Epplen, and J. Schmidtke. 1994. "Alternative reproductive tactics and reproductive success in male rhesus macaques." *Behaviour* 129: 177–201.

Berenbaum, S. A. 2001. "Cognitive function in congenital adrenal hyperplasia." *Endocrinology and Metabolism Clinics of North America* 30: 173–192.

Berger, A. D., J. Satagopan, P. Lee, S. S. Taneja, and I. Osman. 2006. "Differences in clinicopathologic features of prostate cancer between black and white patients treated in the 1990s and 2000s." *Urology* 67: 120–124.

Berger, T. D., and E. Trinkaus. 1995. "Patterns of trauma among the Neandertals." *Journal of Archaeological Science* 22: 841–852.

Berglund, A., and G. Rosenqvist. 2003. "Sex role reversal in pipefish." *Advances in the Study of Behavior* 32: 131–167.

Berling, S., and P. Wolner-Hanssen. 1997. "No evidence of deteriorating semen quality among men in infertile relationships during the last decade: A study of males from southern Sweden." *Human Reproduction* 12: 1002–05.

Berman, M. E., J. I. Tracy, and E. F. Coccaro. 1997. "The serotonin hypothesis of aggression revisited." *Clinical Psychology Review* 17: 651–665.

Bernhardt, P. C., J. M. Dabbs Jr., J. A. Fielden, and C. D. Lutter. 1998. "Testosterone changes during vicarious experiences of winning and losing among fans at sporting events." *Physiology and Behavior* 65: 59–62.

Bernstein, H., G. S. Byers, and R. E. Michod. 1981. "The evolution of sexual reproduction: The importance of DNA repair, complementation, and variation." *American Naturalist* 117: 537–549.

Bersohn, I., and P. J. Oelofse. 1957. "A comparison of urinary oestrogen levels in normal male South African Bantu and European subjects." *South African Medical Journal* 31: 1172–74.

Betzig, L. L., M. B. Mulder, and P. Turke, eds. 1988. *Human Reproductive Be-*

haviour: A Darwinian Perspective. New York: Cambridge University Press.

Bhasin, S., and J. G. Buckwalter. 2001. "Testosterone supplementation in older men: A rational idea whose time has not yet come." *Journal of Andrology* 22: 718–731.

Bhasin, S., T. W. Storer, M. Javanbakht, N. Berman, K. E. Yarasheski, J. Phillips, M. Dike, I. Sinha-Hikim, R. Shen, R. D. Hays, and G. Beall. 2000. "Testosterone replacement and resistance exercise in HIV-infected men with weight loss and low testosterone levels." *Journal of the American Medical Association* 283: 763–770.

Bhasin, S., L. Woodhouse, R. Casaburi, A. B. Singh, R. P. Mac, M. Lee, K. E. Yarasheski, I. Sinha-Hikim, C. Dzekov, J. Dzekov, L. Magliano, and T. W. Storer. 2005. "Older men are as responsive as young men to the anabolic effects of graded doses of testosterone on the skeletal muscle." *Journal of Clinical Endocrinology and Metabolism* 90: 678–688.

Bhasin, S., L. Woodhouse, and T. W. Storer. 2001b. "Proof of the effect of testosterone on skeletal muscle." *Journal of Endocrinology* 170: 27–38.

Bhat, P. N., and A. J. Zavier. 2003. "Fertility decline and gender bias in northern India." *Demography* 40: 637–657.

Bitgood, M. J., L. Shen, and A. P. McMahon. 1996. "Sertoli cell signaling by desert hedgehog regulates the male germline." *Current Biology* 6: 298–304.

Bliege Bird, R. 1999. "Cooperation and conflict: The behavioral ecology of the sexual division of labor." *Evolutionary Anthropology* 8: 65–75.

Bobek, B., K. Perzanowski, and J. Weiner. 1990. "Energy expenditure for reproduction in male red deer." *Journal of Mammalogy* 71(2): 230–232.

Boesch, C. 1991. "The effects of leopard predation on grouping patterns in forest chimpanzees." *Behaviour* 117: 220–242.

Boggs, C. L., and L. E. Gilbert. 1979. "Male contribution to egg production in butterflies: Evidence for transfer of nutrients at mating." *Science* 206: 83–84.

Bogin, B., and R. Keep. 1999. "Eight thousand years of economic and political history in Latin America revealed by anthropometry." *Annals of Human Biology* 26: 333–351.

Bogin, B., M. Wall, and R. B. MacVean. 1992. "Longitudinal analysis of adolescent growth of Ladino and Mayan school children in Guatemala: Effects of environment and sex." *American Journal of Physical Anthropology* 89: 447–457.

Bohossian, H. B., H. Skaletsky, and D. C. Page. 2000. "Unexpectedly similar rates of nucleotide substitution found in male and female hominids." *Nature* 406: 622–625.

Boinski, S. 1987. "Mating patterns in squirrel monkeys *(Saimiri oerstedi):* Implications for seasonal sexual dimorphism." *Behavioral Ecology and Sociobiology* 21: 13–21.

Bolter, D. R., and A. Zihlman. 2003. "Morphometric analysis of growth and development in wild-collected vervet monkeys *(Cercopithecus aethiops),* with implications for growth patterns in Old World monkeys, apes and humans." *Journal of Zoology* (London) 260: 99–110.

Bonsall, R. W., and R. P. Michael. 1992. "Developmental changes in the uptake of testosterone by the primate brain." *Neuroendocrinology* 55: 84–91.

Bonsall, R. W., D. Zumpe, and R. P. Michael. 1990. "Comparisons of the nuclear uptake of [3H]-testosterone and its metabolites by the brains of male and female macaque fetuses at 122 days of gestation." *Neuroendocrinology* 51: 474–480.

Boone, T., and S. Gilmore. 1995. "Effects of sexual intercourse on maximal aerobic power, oxygen pulse, and double product in male sedentary subjects." *Journal of Sports Medicine and Physical Fitness* 35: 214–217.

Boonstra, R., C. J. McColl, and T. J. Karels. 2001. "Reproduction at all costs: The adaptive stress response of male arctic ground squirrels." *Canadian Journal of Zoology* 79: 49–58.

Booth, A., A. C. Mazur, and J. M. Dabbs Jr. 1993. "Endogenous testosterone and competition: The effect of 'fasting.'" *Steroids* 58: 348–350.

Booth, A., G. Shelley, A. Mazur, G. Tharp, and R. Kittok. 1989. "Testosterone, and winning and losing in human competition." *Hormones and Behavior* 23: 556–571.

Borrello, M. E. 2005. "The rise, fall and resurrection of group selection." *Endeavour* 29: 43–47.

Brady, B. M., J. K. Amory, A. Perheentupa, M. Zitzmann, C. J. Hay, D. Apter, R. A. Anderson, W. J. Bremner, P. Pollanen, E. Nieschlag, F. C. Wu, and W. M. Kersemaekers. 2006. "A multicentre study investigating subcutaneous etonogestrel implants with injectable testosterone decanoate as a potential long-acting male contraceptive." *Human Reproduction* 21: 285–294.

Brain, C. K. 1981. *The Hunters or the Hunted? An Introduction to African Cave Taphonomy.* Chicago: University of Chicago Press.

Braun, C. M., I. Montour-Proulx, S. Daigneault, I. Rouleau, S. Kuehn, M. Piskopos, G. Desmarais, F. Lussier, and C. Rainville. 2001. "Prevalence and intellectual outcome of unilateral focal cortical brain damage as a function of age, sex and aetiology." *Behavioural Neurology* 13: 105–116.

Bremner, W. J., M. V. Vitiello, and P. N. Prinz. 1983. "Loss of circadian rhythmicity in blood testosterone levels with aging in normal men." *Journal of Clinical Endocrinology and Metabolism* 56: 1278–81.

Bribiescas, R. G. 1996. "Testosterone levels among Aché hunter/gatherer men: A functional interpretation of population variation among adult males." *Human Nature* 7: 163–188.

—— 1997. "Testosterone as a proximate determinant of somatic energy allocation in human males: Evidence from Ache men of eastern Paraguay." Ph.D. diss., Harvard University.

—— 1998. "Testosterone and dominance: Between-population variance and male energetics." *Brain and Behavioral Sciences* 21: 364–365.

—— 2001a. "Reproductive ecology and life history of the human male." *Yearbook of Physical Anthropology*, 148–176.

—— 2001b. "Reproductive physiology of the human male: An evolutionary and life history perspective." In P. T. Ellison, ed., *Reproductive Ecology and Human Evolution.* New York: Aldine de Gruyter.

—— 2001c. "Serum leptin levels and anthropometric correlates in Ache Amerindians of eastern Paraguay." *American Journal of Physical Anthropology* 115: 297–303.

—— 2005a. "Age-related differences in serum gonadotropin (FSH and LH), salivary testosterone, and 17-beta estradiol levels among Ache Amerindian males of Paraguay." *American Journal of Physical Anthropology* 127: 114–121.

—— 2005b. "Serum leptin levels in Ache Amerindian females with normal adiposity are not significantly different from American anorexia

nervosa patients." *American Journal of Human Biology* 17: 207–
210.

Bribiescas, R. G., and M. S. Hickey. In press. "Population variation and differences in serum leptin independent of adiposity: A comparison of
Ache Amerindian men of Paraguay and lean American male distance
runners." *Nutrition and Metabolism.*

Bromwich, P., J. Cohen, I. Stewart, and A. Walker. 1994. "Decline in sperm
counts: An artefact of changed reference range of 'normal'?" *British
Medical Journal* 309: 19–22.

Brown, A., and G. Blashki. 2005. "Indigenous male health disadvantage—link-
ing the heart and mind." *Australian Family Physician* 34: 813–819.

Brown, D. 1988. "Components of lifetime reproductive success." In *Reproductive Success,* ed. T. H. Clutton-Brock, 439–453. Chicago: University of
Chicago Press.

Brown, F., J. Harris, R. Leakey, and A. Walker. 1985. "Early *Homo erectus* skeleton from west Lake Turkana, Kenya." *Nature* 316: 788–792.

Buchanan, K. L., M. R. Evans, A. R. Goldsmith, D. M. Bryant, and L. V.
Rowe. 2001. "Testosterone influences basal metabolic rate in male
house sparrows: A new cost of dominance signaling?" *Proceedings of
the Royal Society B: Biological Sciences* 268: 1337–44.

Buena, F., R. S. Swerdloff, B. S. Steiner, P. Lutchmansingh, M. A. Peterson,
M. R. Pandian, M. Galmarini, and S. Bhasin. 1993. "Sexual function
does not change when serum testosterone levels are pharmacologi-
cally varied within the normal male range." *Fertility and Sterility* 59:
1118–23.

Buettner, T. 1995. "Sex differentials in old-age mortality." *Population Bulletin
of the United Nations,* 18–44.

Burnham, T. C., J. F. Chapman, P. B. Gray, M. H. McIntyre, S. F. Lipson, and

P. T. Ellison. 2003. "Men in committed, romantic relationships have lower testosterone." *Hormones and Behavior* 44: 119–122.

Buss, D. M., R. J. Larsen, D. Westen, and J. Semmelroth. 1992. "Sex differences in jealousy: Evolution, physiology, and psychology." *Psychological Science* 3: 251–255.

Buss, D. M., and D. P. Schmitt. 1993. "Sexual strategies theory: An evolutionary perspective on human mating." *Psychological Review* 100: 204–232.

Byrne, J., and D. Warburton. 1987. "Male excess among anatomically normal fetuses in spontaneous abortions." *American Journal of Medical Genetics* 26: 605–611.

Campbell, B. C., R. Gillett-Netting, and M. Meloy. 2004. "Timing of reproductive maturation in rural versus urban Tonga boys, Zambia." *Annals of Human Biology* 31: 213–227.

Campbell, B., P. Leslie, J. Quigley, and K. Campbell. 1995. "Hormonal assessment of reproductive function among males in Turkana, Kenya: LH and FSH (Abstract)." *American Journal of Physical Anthropology Supplement* 20: 72.

Campbell, B. C., W. D. Lukas, and K. L. Campbell. 2001. "Reproductive ecology of male immune function and gonadal function." In P. T. Ellison, ed., *Reproductive Ecology and Human Evolution,* 159–178. New York: Aldine de Gruyter.

Campbell, K. L. 1994. "Blood, urine, saliva, and dip-sticks: Experiences in Africa, New Guinea, and Boston." *Annals of the New York Academy of Sciences* 709: 312–330.

Cann, R. L., M. Stoneking, and A. C. Wilson. 1987. "Mitochondrial DNA and human evolution." *Nature* 325: 31–36.

Caretta, N., P. Palego, A. Roverato, R. Selice, A. Ferlin, and C. Foresta. 2006. "Age-matched cavernous peak systolic velocity: A highly sensitive parameter in the diagnosis of arteriogenic erectile dysfunction." *International Journal of Impotence Research* 18: 306–310.

Carlini, A. R., G. A. Daneri, M. E. I. Marquez, G. E. Soave, and S. Poljak. 1997. "Mass transfer from mothers to pups and mass recovery by mothers during the post-breeding foraging period in southern elephant seals *(Mirounga leonina)* at King George Island." *Polar Biology* 18: 305–310.

Carnahan, R. M., and P. J. Perry. 2004. "Depression in aging men: The role of testosterone." *Drugs and Aging* 21: 361–376.

Carruth, L. L., I. Reisert, and A. P. Arnold. 2002. "Sex chromosome genes directly affect brain sexual differentiation." *Nature Neuroscience* 5: 933–934.

Casaburi, R., S. Bhasin, L. Cosentino, J. Porszasz, A. Somfay, M. I. Lewis, M. Fournier, and T. W. Storer. 2004. "Effects of testosterone and resistance training in men with chronic obstructive pulmonary disease." *American Journal of Respiratory and Critical Care Medicine* 170: 870–878.

Chacon-Puignau, G. C., and K. Jaffe. 1996. "Sex ratio at birth deviations in modern Venezuela: The Trivers-Willard effect." *Social Biology* 43: 257–270.

Chapais, B., L. Savard, and C. Gauthier. 2001. "Kin selection and the distribution of altruism in relation to degree of kinship in Japanese macaques *(Macaca fuscata)*." *Behavioral Ecology and Sociobiology* 46: 493–502.

Charnov, E. L. 1993. *Life History Invariants: Some Explorations of Symmetry in Evolutionary Ecology.* New York: Oxford University Press.

Charnov, E. L., and D. Berrigan. 1990. "Dimensionless numbers and life history evolution: Age at maturity versus the adult lifespan." *Evolutionary Ecology* 4: 273–275.

——— 1993. "Why do female primates have such long lifespans and so few babies? or life in the slow lane." Evolutionary Anthropology 1: 191–194.

Chia, S. E., C. N. Ong, L. H. Chua, L. M. Ho, and S. K. Tay. 2000. "Comparison of zinc concentrations in blood and seminal plasma and the various sperm parameters between fertile and infertile men." *Journal of Andrology* 21: 53–57.

Christiansen, K. 1991a. "Androgen and estrogen levels in Namibian Kavango men." *Homo* 42: 43–62.

——— 1991b. "Serum and saliva sex hormone levels in !Kung San men." *American Journal of Physical Anthropology* 86: 37–44.

Clarke, S., R. Kraftsik, H. Van der Loos, and G. M. Innocenti. 1989. "Forms and measures of adult and developing human corpus callosum: Is there sexual dimorphism?" *Journal of Comparative Neurology* 280: 213–230.

Cohen, A. A. 2004. "Female post-reproductive lifespan: A general mammalian trait." *Biological Reviews* 79: 733–750.

Cole, C. J. 2002. "Unisexual clones: Lizards and corals." *Science* 298: 2130–31.

Colfax, J. D. 1990. "The effects of age on levels, pulse characteristics, and circadian rhythm of salivary testosterone." Senior honors (B.A.) thesis, Harvard University.

Coltman, D. W., W. D. Bowen, and J. M. Wright. 1998. "Male mating success in an aquatically mating pinniped, the harbour seal *(Phoca vitulina),* assessed by microsatellite DNA markers." *Molecular Ecology* 7: 627–638.

Cooper, T. G., C. Keck, U. Oberdieck, and E. Nieschlag. 1993. "Effects of multiple ejaculations after extended periods of sexual abstinence on total, motile and normal sperm numbers, as well as accessory gland secretions, from healthy normal and oligozoospermic men." *Human Reproduction* 8: 1251–58.

Couto-Silva, A. C., L. Adan, C. Trivin, and R. Brauner. 2002. "Adult height in advanced puberty with or without gonadotropin hormone releasing hormone analog treatment." *Journal of Pediatric Endocrinology and Metabolism* 15: 297–305.

Couzin, J. 2004. "Research on aging: Gene links calorie deprivation and long life in rodents." *Science* 304: 1731.

Cronk, L. 1991a. "Human behavioral ecology." *Annual Review of Anthropology* 20: 25–53.

—— 1991b. "Wealth, status, and reproductive success among the Mukogodo of Kenya." *American Anthropologist* 93: 345–360.

Crowley, M. A., and K. S. Matt. 1996. "Hormonal regulation of skeletal muscle hypertrophy in rats: The testosterone to cortisol ratio." *European Journal of Applied Physiology* 73: 66–72.

Crowley, W., Jr., R. W. Whitcomb, J. L. Jameson, J. Weiss, J. S. Finkelstein, and L. S. O'Dea. 1991. "Neuroendocrine control of human reproduction in the male." *Recent Progress in Hormone Research* 47: 27–62.

Cumming, D. C., I. A. d. Brunoting, G. Strich, A. L. Ries, and R. W. Rebar. 1986. "Reproductive hormone increases in response to acute exercise in men." *Medicine and Science in Sports and Exercise* 18: 369–373.

Cunningham, G. R., C. M. Ashton, J. F. Annegers, J. Souchek, M. Klima, and B. Miles. 2003. "Familial aggregation of prostate cancer in African-Americans and white Americans." *Prostate* 56: 256–262.

Dabbs, J. M., and M. G. Dabbs. 2000. *Heroes, Rogues, and Lovers: Testosterone and Behavior.* New York: McGraw-Hill.

Daly, M., and M. Wilson. 1983. *Sex, Evolution, and Behavior.* Boston: PWS Publishers.

Darwin, C. 1958. *The Autobiography of Charles Darwin, 1809–1882, With Original Omissions Restored.* Ed. N. Barlow. New York: Harcourt, Brace.

Davatzikos, C., M. Vaillant, S. M. Resnick, J. L. Prince, S. Letovsky, and R. N. Bryan. 1996. "A computerized approach for morphological analysis of the corpus callosum." *Journal of Computer Assisted Tomography* 20: 88–97.

David, K., E. Dingemanse, J. Freud, et al. 1935. "Uber krystallinisches mannliches hormon aus hoden (testosteron) wirksamer als aus harn oder aus cholesterin bereitetes androsteron." *Hoppe-Seyler's Zeitschrift fur Physiologische Chemie* 233: 281.

Davidson, J. M., C. A. Camargo, and E. R. Smith. 1979. "Effects of androgen on sexual behavior in hypogonadal men." *Journal of Clinical Endocrinology and Metabolism* 48: 955–958.

Davies, J. N. P. 1949. "Sex hormone upset in Africans." *British Medical Journal,* 676–679.

de Lacoste, M. C., D. S. Horvath, and D. J. Woodward. 1991. "Possible sex differences in the developing human fetal brain." *Journal of Clinical and Experimental Neuropsychology* 13: 831–846.

de La Rochebrochard, E., J. de Mouzon, F. Thepot, and P. Thonneau. 2006. "Fathers over 40 and increased failure to conceive: The lessons of in vitro fertilization in France." *Fertility and Sterility* 85: 1420–24.

De Rosa, M., S. Zarrilli, A. Di Sarno, N. Milano, M. Gaccione, B. Boggia, G. Lombardi, and A. Colao. 2003. "Hyperprolactinemia in men: Clinical and biochemical features and response to treatment." *Endocrine* 20: 75–82.

De Souza, M. J., and B. E. Miller. 1997. "The effect of endurance training on reproductive function in male runners: A 'volume threshold' hypothesis." *Sports Medicine* 23: 357–374.

de Waal, F. B. M. 2000. *Chimpanzee Politics: Power and Sex among Apes.* Baltimore: Johns Hopkins University Press.

Deacon, T. W. 1997. *The Symbolic Species: The Co-evolution of Language and the Brain.* New York: Norton.

Dean, C., M. G. Leakey, D. Reid, F. Schrenk, G. T. Schwartz, C. Stringer, and A. Walker. 2001. "Growth processes in teeth distinguish modern humans from *Homo erectus* and earlier hominins." *Nature* 414: 628–631.

Dewsbury, D. A. 1982. "Ejaculate cost and male choice." *American Naturalist* 119: 601–610.

Diamond, M. C. 1991. "Hormonal effects on the development of cerebral lateralization." *Psychoneuroendocrinology* 16: 121–129.

Dobzhansky, T. G. 1970. *Genetics of the Evolutionary Process.* New York: Columbia University Press.

Dribe, M. 2004. "Long-term effects of childbearing on mortality: Evidence from pre-industrial Sweden." *Population Studies* (Cambridge) 58: 297–310.

Eaton, N. E., G. K. Reeves, P. N. Appleby, and T. J. Key. 1999. "Endogenous sex hormones and prostate cancer: A quantitative review of prospective studies." *British Journal of Cancer* 80: 930–934.

Edgardh, K. 2002. "Sexual behaviour and early coitarche in a national sample of 17-year-old Swedish boys." *Acta Paediatrica* 91: 985–991.

Edwards, J. H. 1957. "A critical examination of the reputed primary influence of ABO phenotype on fertility and sex ratio." *British Journal of Preventative and Social Medicine* 11: 79–89.

El-Migdadi, F., A. Shotar, Z. El-Akawi, I. Banihani, and R. Abudheese. 2004. "Effect of fasting during the month of Ramadan on serum levels of luteinizing hormone and testosterone in people living in the below sea level environment in the Jordan Valley." *Neuro Endocrinology Letters* 25: 75–77.

Elia, M. 1992. "Organ and tissue contribution to metabolic rate." In J. M. Kinney and H. N. Tucker, eds., *Energy Metabolism: Tissue Determinants and Cellular Corollaries,* 61–79. New York: Raven Press.

Ellis, L., and H. Nyborg. 1992. "Racial/ethnic variations in male testosterone levels: A probable contributor to group differences in health." *Steroids* 57: 72–75.

Ellison, P. T. 1988. "Human salivary steroids: Methodological considerations and applications in physical anthropology." *Yearbook of Physical Anthropology* 31: 115–142.

——— 1996. "Developmental influences on adult ovarian function." *American Journal of Human Biology* 8: 725–734.

——— 2001. *On Fertile Ground: A Natural History of Human Reproduction.* Cambridge, Mass.: Harvard University Press.

——— 2003. "Energetics and reproductive effort." *American Journal of Human Biology* 15: 342–351.

Ellison, P. T., R. G. Bribiescas, G. R. Bentley, B. C. Campbell, S. F. Lipson, C. Panter-Brick, and K. R. Hill. 2002. "Population variation in age-related decline in male salivary testosterone." *Human Reproduction* 17: 3251-53.

Ellison, P. T., and C. Lager. 1986. "Moderate recreational running is associated with lowered salivary progesterone profiles in women." *American Journal of Obstetrics and Gynecology* 154: 1000-03.

Ellison, P. T., S. F. Lipson, and M. D. Meredith. 1989a. "Salivary testosterone levels in males from the Ituri forest of Zaire." *American Journal of Human Biology* 1: 21-24.

Ellison, P. T., and C. Panter-Brick. 1996. "Salivary testosterone levels among Tamang and Kami males of central Nepal." *Human Biology* 68: 955-965.

Ellison, P. T., N. R. Peacock, and C. Lager. 1989b. "Ecology and ovarian function among Lese women of the Ituri Forest, Zaire." *American Journal of Physical Anthropology* 78: 519-526.

Emanuel, E. R., E. T. Goluboff, and H. Fisch. 1998. "MacLeod revisited: Sperm count distributions in 374 fertile men from 1971 to 1994." *Urology* 51: 86-88.

Emlen, D. J. 1997. "Alternative reproductive tactics and male-dimorphism in the horned beetle *Onthophagus acumunatus* (Coleoptera: Scarabaeidae)." *Behavioral Ecology and Sociobiology* 41: 335-341.

Ewald, P. W. 2004. "Evolution of virulence." *Infectious Disease Clinics of North America* 18: 1-15.

Feigelson, M., and D. M. Linkie. 1987. "Effects of androgen administered to neonatal male rats on hypophyseal gonadotropin content, secretion and response to LHRH at adulthood." *Hormone Research* 27: 159-167.

Fernandez-Guasti, A., F. P. Kruijver, M. Fodor, and D. F. Swaab. 2000. "Sex differences in the distribution of androgen receptors in the human hypothalamus." *Journal of Comparative Neurology* 425: 422–435.

Finch, C. E., and M. R. Rose. 1995. "Hormones and the physiological architecture of life history evolution." *Quarterly Review of Biology* 70: 1–52.

Fisch, H., and E. T. Goluboff. 1996. "Geographic variations in sperm counts: A potential cause of bias in studies of semen quality." *Fertility and Sterility* 65: 1044–46.

Fisch, H., E. F. Ikeguchi, and E. T. Goluboff. 1996. "Worldwide variations in sperm counts." *Urology* 48: 909–911.

Fisher, D. C. 1985. "Evolutionary morphology: Beyond the analogous, the anecdotal, and the ad hoc." *Paleobiology* 11: 120–138.

Fisher, R. A. 1930. *The Genetical Theory of Natural Selection.* New York: Dover.

Fitzpatrick, J. M., R. S. Kirby, R. J. Krane, J. Adolfsson, D. W. Newling, and I. Goldstein. 1998. "Sexual dysfunction associated with the management of prostate cancer." *European Urology* 33: 513–522.

Flinn, M. V. 1992. "Paternal care in a Caribbean village." In Hewlett 1992, 57–84.

Flinn, M. V., and B. G. England. 1997. "Social economics of childhood glucocorticoid stress response and health." *American Journal of Physical Anthropology* 102: 33–53.

Folstad, I., and A. K. Karter. 1992. "Parasites, bright males and the immunocompetence handicap." *American Naturalist* 139: 603–622.

Forger, N. G., B. G. Galef Jr., and M. M. Clark. 1996. "Intrauterine position affects motoneuron number and muscle size in a sexually dimorphic neuromuscular system." *Brain Research* 735: 119–124.

Fowler, G. S., J. C. Wingfield, P. D. Boersma, and R. A. Sosa. 1994. "Reproductive endocrinology and weight change in relation to reproductive success in the magellanic penguin *(Spheniscus magellanicus)*." *General and Comparative Endocrinology* 94: 305–315.

Fox, M. S., and R. A. Reijo Pera. 2002. "Male infertility: Genetic analysis of the DAZ genes on the human Y chromosome and genetic analysis of DNA repair." *Molecular and Cellular Endocrinology* 186: 231–239.

Francis, C. M., E. L. P. Anthony, J. A. Brunton, and T. H. Kunz. 1994. "Lactation in male fruit bats." *Nature* 367: 691–692.

Freeman, E. R., D. A. Bloom, and E. J. McGuire. 2001. "A brief history of testosterone." *Journal of Urology* 165: 371–373.

Fremon, C. 1995. *Father Greg and the Homeboys: The Extraordinary Journey of Father Greg Boyle and His Work with the Latino Gangs of East L.A.* New York: Hyperion.

Fukuda, M., K. Fukuda, T. Shimizu, and H. Moller. 1998. "Decline in sex ratio at birth after Kobe earthquake." *Human Reproduction* 13: 2321–22.

Fuse, H., T. Kazama, S. Ohta, and Y. Fujiuchi. 1999. "Relationship between zinc concentrations in seminal plasma and various sperm parameters." *International Urology and Nephrology* 31: 401–408.

Futuyma, D. J. 1986. *Evolutionary Biology.* Sunderland, Mass.: Sinauer Associates.

Galard, R., M. Antolin, R. Catalan, P. Magana, S. Schwartz, and J. M. Castellanos. 1987. "Salivary testosterone levels in infertile men." *International Journal of Andrology* 10: 597–601.

Galdas, P. M., F. Cheater, and P. Marshall. 2005. "Men and health help-seeking behaviour: Literature review." *Journal of Advanced Nursing* 49: 616–623.

Ganatra, B., and S. Hirve. 1994. "Male bias in health care utilization for un-
der-fives in a rural community in western India." *Bulletin of the World
Health Organization* 72: 101–104.

Gann, P. H., C. H. Hennekens, J. Ma, C. Longcope, and M. J. Stampfer. 1996.
"Prospective study of sex hormone levels and risk of prostate cancer."
Journal of the National Cancer Institute 88: 1118–26.

Gaulin, S. J., and C. J. Robbins. 1991. "Trivers-Willard effect in contemporary
North American society." *American Journal of Physical Anthropology*
85: 61–69.

Geary, M. P., P. J. Pringle, C. H. Rodeck, J. C. Kingdom, and P. C. Hindmarsh. 2003. "Sexual dimorphism in the growth hormone and insulin-like growth factor axis at birth." *Journal of Clinical Endocrinology
and Metabolism* 88: 3708–14.

George, D. T., J. C. Umhau, M. J. Phillips, D. Emmela, P. W. Ragan, S. E.
Shoaf, and R. R. Rawlings. 2001. "Serotonin, testosterone and alcohol
in the etiology of domestic violence." *Psychiatry Research* 104: 27–37.

Gibson, M. A., and R. Mace. 2003. "Strong mothers bear more sons in rural
Ethiopia." Proceedings. *Biological Sciences* 270 Supplement 1: S108–
109.

Goldizen, A. W. 1987. "Tamarins and marmosets: Communal care of offspring." In B. B. Smuts, D. L. Cheney, R. M. Seyfarth, R. W.
Wrangham, and T. T. Struhsaker, eds., *Primate Societies,* 34–43. Chicago: University of Chicago Press.

Goodman, F. R., and P. J. Scambler. 2001. "Human HOX gene mutations."
Clinical Genetics 59: 1–11.

Gorman-Stapleton, O. 1990. "Prohibiting amniocentesis in India: A solution
to the problem of female infanticide or a problem to the solution of
prenatal diagnosis?" *ILSA Journal of International Law* 14: 23–43.

Gould, K. G., M. Flint, and C. E. Graham. 1981. "Chimpanzee reproductive senescence: A possible model for evolution of the menopause." *Maturitas* 3: 157–166.

Gould, S. J. 1981. *The Mismeasure of Man.* New York: Norton.

Graafmans, W. C., J. H. Richardus, A. Macfarlane, M. Rebagliato, B. Blondel, S. P. Verloove-Vanhorick, and J. P. Mackenbach. 2001. "Comparability of published perinatal mortality rates in Western Europe: The quantitative impact of differences in gestational age and birthweight criteria." *BJOG: An International Journal of Obstetrics and Gynaecology* 108: 1237–45.

Graffelman, J., E. F. Fugger, K. Keyvanfar, and J. D. Schulman. 1999. "Human live birth and sperm-sex ratios compared." *Human Reproduction* 14: 2917–20.

Gray, A., H. A. Feldman, J. B. McKinlay, and C. Longcope. 1991. "Age, disease, and changing sex hormone levels in middle-aged men: Results of the Massachusetts Male Aging Study." *Journal of Clinical Endocrinology and Metabolism* 73: 1016–25.

Gray, P. B. 2003. "Marriage, parenting, and testosterone variation among Kenyan Swahili men." *American Journal of Physical Anthropology* 122: 279–286.

Gray, P., and B. Campbell. 2005. "Erectile dysfunction and its correlates among the Ariaal of northern Kenya." *International Journal of Impotence Research* 17: 445–449.

Gray, P. B., S. M. Kahlenberg, E. S. Barrett, S. F. Lipson, and P. T. Ellison. 2002. "Marriage and fatherhood are associated with lower testosterone in males." *Evolution and Human Behavior* 23: 193–201.

Grimshaw, G. M., M. P. Bryden, and J. K. Finegan. 1995. "Relations between prenatal testosterone and cerebral lateralization in children." *Neuropsychology* 9: 68–79.

Grossman, C. 1989. "Possible underlying mechanisms of sexual dimorphism in the immune response, fact and hypothesis." *Journal of Steroid Biochemistry* 34: 241–251.

Guinness Media. 1996. *Guinness Book of World Records.* Stamford, CT.

Guzick, D. S., J. W. Overstreet, P. Factor-Litvak, C. K. Brazil, S. T. Nakajima, C. Coutifaris, S. A. Carson, P. Cisneros, M. P. Steinkampf, J. A. Hill, D. Xu, and D. L. Vogel. 2001. "Sperm morphology, motility, and concentration in fertile and infertile men." *New England Journal of Medicine* 345: 1388–93.

Gwynne, D. T. 1981. "Sexual difference theory: Mormon crickets show role reversal in mate choice." *Science* 213: 779–780.

Haavisto, A. M., K. Pettersson, M. Bergendahl, A. Virkamaki, and I. Huhtaniemi. 1995. "Occurrence and biological properties of a common genetic variant of luteinizing hormone." *Journal of Clinical Endocrinology and Metabolism* 80: 1257–1263.

Haig, D. 1993. "Genetic conflicts in human pregnancy." *Quarterly Review of Biology* 68: 495–532.

Haldane, J. B. S. 1947. "The mutation rate of the gene for haemophilia, and its segregation ratios in males and females." *Annals of Eugenics* 13: 262–271.

Hall, H. L., M. G. Flynn, K. K. Carroll, P. G. Brolinson, S. Shapiro, and B. A. Bushman. 1999. "Effects of intensified training and detraining on testicular function." *Clinical Journal of Sport Medicine* 9: 203–208.

Hamilton, W. D. 1964. "The genetical evolution of social behaviour. I and II." *Journal of Theoretical Biology* 7: 1–52.

Hamilton, W. D., and M. Zuk. 1982. "Heritable true fitness and bright birds: A role for parasites?" *Science* 218: 384–387.

Harman, S. M., E. J. Metter, J. D. Tobin, J. Pearson, and M. R. Blackman. 2001. "Longitudinal effects of aging on serum total and free testosterone levels in healthy men: Baltimore Longitudinal Study of Aging." *Journal of Clinical Endocrinology and Metabolism* 86: 724–731.

Harvey, P. H., and M. D. Pagel. 1991. *The Comparative Method in Evolutionary Biology.* New York: Oxford University Press.

Hassold, T., S. D. Quillen, and J. A. Yamane. 1983. "Sex ratio in spontaneous abortions." *Annals of Human Genetics* 47, pt. 1: 39–47.

Hausfater, G., and S. B. Hrdy. 1984. *Infanticide: Comparative and Evolutionary Perspectives.* New York: Aldine.

Hawkes, K., and R. B. Bird. 2002. "Showing off, handicap signaling, and the evolution of men's work." *Evolutionary Anthropology* 11: 58–67.

Hawkes, K., J. F. O' Connell, and N. G. Blurton-Jones. 1997. "Hadza women's time allocation, offspring provisioning, and the evolution of long postmenopausal life spans." *Current Anthropology* 38: 551–577.

Hawkes, K., A. R. Rogers, and E. L. Charnov. 1995. "The male's dilemma: Increased offspring production is more paternity to steal." *Evolutionary Ecology* 9: 662–677.

Hayes, F. J., S. B. Seminara, S. Decruz, P. A. Boepple, and W. F. Crowley Jr. 2000. "Aromatase inhibition in the human male reveals a hypothalamic site of estrogen feedback." *Journal of Clinical Endocrinology and Metabolism* 85: 3027–35.

Heald, A. H., F. Ivison, S. G. Anderson, K. Cruickshank, I. Laing, and J. M. Gibson. 2003. "Significant ethnic variation in total and free testosterone concentration." *Clinical Endocrinology* (Oxford) 58: 262–266.

Helle, S., V. Lummaa, and J. Jokela. 2002. "Sons reduced maternal longevity in preindustrial humans." *Science* 296: 1085.

Henriksson, J. 1992. "Energy metabolism in muscle: Its possible role in the adaptation to energy deficiency." In J. M. Kinney and H. N. Tucker, eds., *Energy Metabolism: Tissue Determinants and Cellular Corollaries,* 345–365. New York: Raven Press.

Herbst, K. L., and S. Bhasin. 2004. "Testosterone action on skeletal muscle." *Current Opinion in Clinical Nutrition and Metabolic Care* 7: 271–277.

Hering-Hanit, R., R. Achiron, S. Lipitz, and A. Achiron. 2001. "Asymmetry of fetal cerebral hemispheres: In utero ultrasound study." *Archives of Disease in Childhood: Fetal and Neonatal Edition* 85: F194–196.

Hewlett, B. S., ed. 1992. *Father-Child Relations: Cultural and Biosocial Contexts,* Hawthorne, N.Y.: Aldine De Gruyter.

Hickey, M. S., R. V. Considine, R. G. Israel, T. L. Mahar, M. R. McCammon, G. L. Tyndall, J. A. Houmard, and J. F. Caro. 1996. "Leptin is related to body fat content in male distance runners." *American Journal of Physiology* 271: E938–940.

Hill, K. R. 1993. "Life history theory and evolutionary anthropology." *Evolutionary Anthropology* 2: 78–88.

Hill, K. R., C. Boesch, J. Goodall, A. Pusey, J. Williams, and R. W. Wrangham. 2001. "Mortality rates among wild chimpanzees." *Journal of Human Evolution* 40: 437–450.

Hill, K. R., and A. M. Hurtado. 1992. "Paternal effect on offspring survivorship among Ache and Hiwi hunter-gatherers: Implications for modeling pair-bond stability." In Hewlett 1992, 31–56.

——— 1996. *Ache Life History: The Ecology and Demography of a Foraging People*. New York: Aldine de Gruyter.

Hill, K. R., H. Kaplan, K. Hawkes, and A. M. Hurtado. 1985. "Men's time allocation to subsistence work among the Ache of eastern Paraguay." *Human Ecology* 13: 29–47.

Hindmarsh, P. C., M. P. Geary, C. H. Rodeck, J. C. Kingdom, and T. J. Cole. 2002. "Intrauterine growth and its relationship to size and shape at birth." *Pediatric Research* 52: 263–268.

Hines, M., and R. A. Gorski. 1985. "The dual brain: Hemispheric specialization." In F. Benson and E. Zaidel, eds., *Humans,* 75–96. New York: Guilford.

Hines, M., and C. Shipley. 1984. "Prenatal exposure to diethylstilbestrol (DES) and the development of sexually dimorphic cognitive abilities and cerebral lateralization." *Developmental Psychology* 20: 81–94.

Hirota, T., and Y. Obara. 2000. "Time allocation to the reproductive and feeding behaviors in the male cabbage butterfly." *Zoological Science* (Tokyo) 17: 323–327.

Hoberman, J. M., and C. E. Yesalis. 1995. "The history of synthetic testosterone." *Scientific American* 272: 76–81.

Hoffer, L. J., I. Z. Beitins, N. H. Kyung, and B. R. Bistrian. 1986. "Effects of severe dietary restriction on male reproductive hormones." *Journal of Clinical Endocrinology and Metabolism* 62: 288–292.

Holloway, R. L., and M. C. de Lacoste. 1986. "Sexual dimorphism in the human corpus callosum: An extension and replication study." *Human Neurobiology* 5: 87–91.

Howell, N. 1979. *Demography of the Dobe !Kung*. New York: Academic Press.

Howie, B. J., and T. D. Schultz. 1985. "Dietary and hormonal interrelationships among vegetarian Seventh-Day Adventists and nonvegetarian men." *American Journal of Clinical Nutrition* 42: 127–134.

Hrdy, S. 1981. *The Woman That Never Evolved.* Cambridge, Mass.: Harvard University Press.

——— 1999. *Mother Nature: A History of Mothers, Infants, and Natural Selection.* New York: Pantheon.

Huang, H. Y., C. L. Lee, Y. M. Lai, M. Y. Chang, H S. Wang, S. Y. Chang, and Y. K. Soong. 1996. "The impact of the total motile sperm count on the success of intrauterine insemination with husband's spermatozoa." *Journal of Assisted Reproduction and Genetics* 13: 56–63.

Hudson, V. M., and A. M. den Boer. 2004. *Bare Branches: The Security Implications of Asia's Surplus Male Population.* Cambridge, Mass.: MIT Press.

Huhtaniemi, I. T. 2002a. "LH and FSH receptor mutations and their effects on puberty." *Hormone Research* 57 Supplement 2: 35–38.

——— 2002b. "The role of mutations affecting gonadotrophin secretion and action in disorders of pubertal development." *Best Practice and Research: Clinical Endocrinology and Metabolism* 16: 123–138.

Huhtaniemi, I. T., and K. S. Pettersson. 1999. "Alterations in gonadal steroidogenesis in individuals expressing a common genetic variant of luteinizing hormone." *Journal of Steroid Biochemistry and Molecular Biology* 69: 281–285.

Hulme, H. B. 1951. "Effect of semistarvation on human semen." *Fertility and Sterility* 2: 319–331.

Humphries, S., and D. J. Stevens. 2001. "Reproductive biology: Out with a bang." *Nature* 410: 758–759.

Hurtado, A. M., and K. R. Hill. 1987. "Early dry season subsistence ecology of Cuiva (Hiwi) foragers of Venezuela." *Human Ecology* 15: 163–187.

Hwang, S. J., E. K. Ji, E. K. Lee, Y. M. Kim, Y. Shin da, Y. H. Cheon, and I. J. Rhyu. 2004. "Gender differences in the corpus callosum of neonates." *Neuroreport* 15: 1029–32.

Imperato-McGinley, J., R. E. Peterson, T. Gautier, and E. Sturla. 1979. "Androgens and the evolution of male-gender identity among male pseudohermaphrodites with 5 alpha-reductase deficiency." *New England Journal of Medicine* 300: 1233–1237.

Ingemarsson, I. 2003. "Gender aspects of preterm birth." *BJOG: An International Journal of Obstetrics and Gynaecology* 110 Supplement 20: 34–38.

Iribarren, C., D. S. Sharp, C. M. Burchfiel, and H. Petrovitch. 1995. "Association of weight loss and weight fluctuation with mortality among Japanese American men." *New England Journal of Medicine* 333: 686–692.

Irving, J., A. Bittles, J. Peverall, A. Murch, and P. Matson. 1999. "The ratio of X- and Y-bearing sperm in ejaculates of men with three or more children of the same sex." *Journal of Assisted Reproduction and Genetics* 16: 492–494.

Isaacs, J. T. 1996. "Role of androgens in normal and malignant growth of the prostate." In S. Bhasin, H. L. Gabelnick, J. M. Spieler, R. S. Swerdloff, C. Wang, and C. Kelly, eds., *Pharmacology, Biology, and Clinical Applications of Androgens,* 95–102. New York: Wiley-Liss.

Jacklin, C. N., E. E. Maccoby, C. H. Doering, and D. R. King. 1984. "Neonatal sex-steroid hormones and muscular strength of boys and girls in the first three years." *Developmental Psychobiology* 17: 301–310.

Jameson, J. L. 1996. "Inherited disorders of the gonadotropin hormones." *Molecular and Cellular Endocrinology* 125: 143–149.

Jamison, C. S., L. L. Cornell, P. L. Jamison, and H. Nakazato. 2002. "Are all grandmothers equal? A review and a preliminary test of the 'Grandmother Hypothesis' in Tokugawa Japan." *American Journal of Physical Anthropology* 119: 67–76.

Jasienska, G., and P. T. Ellison. 1998. "Physical work causes suppression of ovarian function in women." *Proceedings of the Royal Society of London: Series B* 265: 1847–51.

Jasienska, G., and I. Thune. 2001. "Lifestyle, hormones, and risk of breast cancer." *British Medical Journal* 322: 586–587.

Jasienska, G., I. Thune, and P. T. Ellison. 2000. "Energetic factors, ovarian steroids and the risk of breast cancer." *European Journal of Cancer Prevention* 9: 231–239.

Jathar, V. S., R. Hirwe, S. Desai, and R. S. Satoskar. 1976. "Dietetic habits and quality of semen in Indian subjects." *Andrologia* 8: 355–358.

Jean-Faucher, C., M. Berger, M. de Turckheim, G. Veyssiere, and C. Jean. 1982. "The effect of preweaning undernutrition upon the sexual development of male mice." *Biology of the Neonate* 41: 45–51.

Jennions, M. D., and P. R. Y. Backwell. 1996. "Residency and size affect fight duration and outcome in the fiddler crab *Uca annulipes*." *Biological Journal of the Linnaean Society* 57: 292–306.

—— 1998. "Variation in courtship rate in the fiddler crab *Uca annulipes:* Is it related to male attractiveness?" *Behavioral Ecology* 9: 605–611.

Ji, W., M. N. Clout, and S. D. Sarre. 2000. "Responses of male brushtail possums to sterile females: Implications for biological control." *Journal of Applied Ecology* 37: 926–934.

Jin, B., J. Beilin, J. Zajac, and D. J. Handelsman. 2000. "Androgen receptor gene polymorphism and prostate zonal volumes in Australian and Chinese men." *Journal of Andrology* 21: 91–98.

Kaiser, F. E. 1999. "Erectile dysfunction in the aging man." *Medical Clinics of North America* 83: 1267–78.

Kaplan, H., and H. Dove. 1987. "Infant development among the Ache of eastern Paraguay." *Developmental Psychology* 23: 190–198.

Kaplan, H., and K. R. Hill. 1985. "Hunting ability and reproductive success among male Ache foragers: Preliminary results." *Current Anthropology* 26: 131–133.

Kaplan, H., K. R. Hill, J. Lancaster, and A. M. Hurtado. 2000. "A theory of human life history evolution: Diet, intelligence and longevity." *Evolutionary Anthropology* 9: 156–184.

Kaplowitz, P. 2004. "Precocious puberty: Update on secular trends, definitions, diagnosis, and treatment." *Advances in Pediatrics* 51: 37–62.

Kellokumpu-Lehtinen, P., and L. J. Pelliniemi. 1984. "Sex ratio of human conceptuses." *Obstetrics and Gynecology* 64: 220–222.

Kenny, A. M., S. Bellantonio, C. A. Gruman, R. D. Acosta, and K. M. Prestwood. 2002. "Effects of transdermal testosterone on cognitive function and health perception in older men with low bioavailable testosterone levels." *Journals of Gerontology, Series A, Biological Sciences and Medical Sciences* 57: M321–325.

Ketterson, E. D., and V. Nolan Jr. 1992. "Hormones and life histories: An integrative approach." *American Naturalist* 140 supplement: S33–S62.

Ketterson, E. D., V. Nolan Jr., L. Wolf, and C. Ziegenfus. 1992. "Testosterone and avian life histories: Effects of experimentally elevated testosterone on behavior and correlates of fitness in the dark-eyed junco (*Junco hyemalis*)." *American Naturalist* 140: 980–999.

Key, T. 1995. "Risk factors for prostate cancer." *Cancer Surveys* 23: 63–77.

King, J. C. 2003. "The risk of maternal nutritional depletion and poor outcomes increases in early or closely spaced pregnancies." *Journal of Nutrition* 133: 1732S–36S.

Kinsey, A. C., W. B. Pomeroy, and C. E. Martin. 1948. *Sexual Behavior in the Human Male.* Philadelphia: W. B. Saunders.

Kirkwood, T. B. 1977. "Evolution of ageing." *Nature* 270: 301–304.

——— 1999. *Time of Our Lives: The Science of Human Aging.* Oxford; New York: Oxford University Press.

Kivlighan, K. T., D. A. Granger, and A. Booth. 2005. "Gender differences in testosterone and cortisol response to competition." *Psychoneuroendocrinology* 30: 58–71.

Klein, S. L. 2000. "The effects of hormones on sex differences in infection: From genes to behavior." *Neuroscience and Biobehavioral Reviews* 24: 627–638.

——— 2004. "Hormonal and immunological mechanisms mediating sex differences in parasite infection." *Parasite Immunology* 26: 247–264.

Klibanski, A., I. Z. Beitins, T. Badger, R. Little, and J. W. McArthur. 1981. "Reproductive function during fasting in men." *Journal of Clinical Endocrinology and Metabolism* 53: 258–263.

Kline, J., and B. Levin. 1992. "Trisomy and age at menopause: Predicted associations given a link with rate of oocyte atresia." *Paediatric and Perinatal Epidemiology* 6: 225–239.

Ko, K. T., T. E. Needham, and H. Zia. 1998. "Emulsion formulations of testosterone for nasal administration." *Journal of Microencapsulation* 15: 197–205.

Koga, T., P. R. Y. Backwell, J. H. Christy, M. Murai, and E. Kasuya. 2001. "Male-biased predation of a fiddler crab." *Animal Behaviour* 62: 201–207.

Koga, T., M. Murai, and H. S. Yong. 1999. "Male-male competition and intersexual interactions in underground mating of the fiddler crab *Uca paradussumieri.*" *Behaviour* 136: 651–667.

Kohut, M. L., J. R. Thompson, J. Campbell, G. A. Brown, M. D. Vukovich, D. A. Jackson, and D. S. King. 2003. "Ingestion of a dietary supplement containing dehydroepiandrosterone (DHEA) and androstenedione has minimal effect on immune function in middle-aged men." *Journal of the American College of Nutrition* 22: 363–371.

Konner, M. 1976. "Maternal care, infant behavior and development among the !Kung." In R. B. Lee and I. DeVore, eds., *Kalahari Hunter-Gatherers: Studies of the !Kung San and Their Neighbors,* 218–245. Cambridge, Mass.: Harvard University Press.

Konner, M., and C. Worthman. 1980. "Nursing frequency, gonadal function, and birth spacing among !Kung hunter-gatherers." *Science* 207: 788–791.

Koopman, P., J. Gubbay, P. Vivian, R. Goodfellow, and R. Lovell-Badge. 1991. "Male development of chromosomally female mice transgenic for *Sry.*" *Nature* 351: 117–121.

Koziel, S., and S. J. Ulijaszek. 2001. "Waiting for Trivers and Willard: Do the rich really favor sons?" *American Journal of Physical Anthropology* 115: 71–79.

Kratzik, C. W., G. Schatzl, G. Lunglmayr, E. Rucklinger, and J. Huber. 2005.

"The impact of age, body mass index and testosterone on erectile dysfunction." *Journal of Urology* 174: 240–243.

Krucken, J., M. A. Dkhil, J. V. Braun, R. M. Schroetel, M. El-Khadragy, P. Carmeliet, H. Mossmann, and F. Wunderlich. 2005. "Testosterone suppresses protective responses of the liver to blood-stage malaria." *Infection and Immunity* 73: 436–443.

Kruger, D. J., and R. M. Nesse. 2002. "Sexual selection and the male:female mortality ratio." *Evolutionary Psychology* 2: 66–85.

Kuhnert, B., M. Byrne, M. Simoni, W. Kopcke, J. Gerss, G. Lemmnitz, and E. Nieschlag. 2005. "Testosterone substitution with a new transdermal, hydroalcoholic gel applied to scrotal or non-scrotal skin: A multicentre trial." *European Journal of Endocrinology* 153: 317–326.

Kulynych, J. J., K. Vladar, D. W. Jones, and D. R. Weinberger. 1994. "Gender differences in the normal lateralization of the supratemporal cortex: MRI surface-rendering morphometry of Heschl's gyrus and the planum temporale." *Cerebral Cortex* 4: 107–118.

Kumar, T. R., Y. Wang, N. Lu, and M. M. Matzuk. 1997. "Follicle stimulating hormone is required for ovarian follicle maturation but not male fertility." *Nature Genetics* 15: 201–204.

Kurihara, M., K. Aoki, and S. Hisamichi, eds. 1989. *Cancer Mortality Statistics in the World* 1950–1985. Nagoya: University of Nagoya Press.

Kuzawa, C. W. 2005. "Fetal origins of developmental plasticity: Are fetal cues reliable predictors of future nutritional environments?" *American Journal of Human Biology* 17: 5–21.

La Cava, A., and G. Matarese. 2004. "The weight of leptin in immunity." *Nature Reviews: Immunology* 4: 371–379.

Lack, D., J. A. Gibb, and D. F. Owen. 1957. "Survival in relation to brood-size in tits." *Proceedings of the Zoological Society of London* 128: 313–326.

Laden, G., and R. Wrangham. 2005. "The rise of the hominids as an adaptive shift in fallback foods: Plant underground storage organs (USOs) and australopith origins." *Journal of Human Evolution* 49: 482–498.

Lager, C., and P. T. Ellison. 1987. "Effects of moderate weight loss on ovulatory frequency and luteal function in adult women." *American Journal of Physical Anthropology* 72: 221–222.

—— 1990. "Effect of moderate weight loss on ovarian function assessed by salivary progesterone measurements." *American Journal of Human Biology* 2: 303–312.

Larme, A. C. 1997. "Health care allocation and selective neglect in rural Peru." *Social Science and Medicine* 44: 1711–23.

Layman, L. C., E. J. Lee, D. B. Peak, A. B. Namnoum, K. V. Vu, B. L. van Lingen, M. R. Gray, P. G. McDonough, R. H. Reindollar, and J. L. Jameson. 1997. "Delayed puberty and hypogonadism caused by mutations in the follicle-stimulating hormone beta-subunit gene." *New England Journal of Medicine* 337: 607–611.

Layman, L. C., and P. G. McDonough. 2000. "Mutations of follicle stimulating hormone-beta and its receptor in human and mouse: Genotype/ phenotype." *Molecular and Cellular Endocrinology* 161: 9–17.

Lazar, L., A. Pertzelan, N. Weintrob, M. Phillip, and R. Kauli. 2001. "Sexual precocity in boys: Accelerated versus slowly progressive puberty gonadotropin-suppressive therapy and final height." *Journal of Clinical Endocrinology and Metabolism* 86: 4127–32.

Le Boeuf, B. J. 1974. "Male-male competition and reproductive success in elephant seals." *American Zoologist* 14: 163–176.

Leigh, S. R. 1995. "Socioecology and the ontogeny of sexual size dimorphism in anthropoid primates." *American Journal of Physical Anthropology* 97: 339–356.

——— 1996. "Evolution of human growth spurts." *American Journal of Physical Anthropology* 101: 455–474.

Leonard, W. R., and M. L. Robertson. 1997. "Comparative primate energetics and hominid evolution." *American Journal of Physical Anthropology* 102: 265–281.

Lin, S., Q. F. Xing, and Y. W. Chien. 1999. "Transdermal testosterone delivery: Comparison between scrotal and nonscrotal delivery systems." *Pharmaceutical Development and Technology* 4: 405–414.

Lipson, S. F., and P. T. Ellison. 1989. "Development of the protocols for the application of salivary steroid analyses to field conditions." *American Journal of Human Biology* 1: 249–255.

——— 1996. "Comparison of salivary steroid profiles in naturally occurring conception and non-conception cycles." *Human Reproduction* 11: 2090–96.

Lockwood, C. A. 1999. "Sexual dimorphism in the face of *Australopithecus africanus*." *American Journal of Physical Anthropology* 108: 97–127.

Lord, G. M., G. Matarese, J. K. Howard, R. J. Baker, S. R. Bloom, and R. I. Lechler. 1998. "Leptin modulates the T-cell immune response and reverses starvation-induced immunosuppression." *Nature* 394: 897–901.

Love, O. P., E. H. Chin, K. E. Wynne-Edwards, and T. D. Williams. 2005. "Stress hormones: A link between maternal condition and sex-biased reproductive investment." *American Naturalist* 166(6): 751–66.

Lovejoy, C. O. 1981. "The origin of man." *Science* 211: 341–350.

Lummaa, V., J. Merila, and A. Kause. 1998. "Adaptive sex ratio variation in pre-industrial human *(Homo sapiens)* populations?" *Proceedings of the Royal Society of London. Series B* 265: 563–568.

Luukkaa, V., U. Pesonen, I. Huhtaniemi, A. Lehtonen, R. Tilvis, J. Tuomilehto, M. Koulu, and R. Huupponen. 1998. "Inverse correlation between serum testosterone and leptin in men." *Journal of Clinical Endocrinology and Metabolism* 83: 3243–46.

Luukkaa, V., J. Rouru, O. Ahokoski, H. Scheinin, K. Irjala, and R. Huupponen. 2000. "Acute inhibition of oestrogen biosynthesis does not affect serum leptin levels in young men." *European Journal of Endocrinology* 142: 164–169.

Lyell, C. 1842. *Principles of Geology: or, The Modern Changes of the Earth and its Inhabitants, Considered as Illustrative of Geology.* Boston: Hilliard Gray.

Ma, Z., R. L. Gingerich, J. V. Santiago, S. Klein, C. H. Smith, and M. Landt. 1996. "Radioimmunoassay of leptin in human plasma." *Clinical Chemistry* 42: 942–946.

Mace, R. 1996. "Biased parental investment and reproductive success in Gabbra pastoralists." *Behavioral Ecology and Sociobiology* 38: 75–81.

—— 2000. "Evolutionary ecology of human life history." *Animal Behaviour* 59: 1–10.

Mace, R., F. Jordan, and C. Holden. 2003. "Testing evolutionary hypotheses about human biological adaptation using cross-cultural comparison." *Comparative Biochemistry and Physiology, Part A, Molecular and Integrative Physiology* 136: 85–94.

Maconochie, N., and E. Roman. 1997. "Sex ratios: Are there natural variations within the human population?" *British Journal of Obstetrics and Gynaecology* 104: 1050–53.

Maffei, M., J. Halaas, E. Ravussin, R. E. Pratley, G. H. Lee, Y. Zhang, H. Fei, S. K. R., R. Lallone, S. Ranganathan, et al. 1995. "Leptin levels in human and rodent: Measurement of plasma leptin and ob RNA in obese and weight-reduced subjects." *Nature Medicine* 1: 1155–61.

Main, K. M., I. M. Schmidt, and N. E. Skakkebaek. 2000. "A possible role for reproductive hormones in newborn boys: Progressive hypogonadism without the postnatal testosterone peak." *Journal of Clinical Endocrinology and Metabolism* 85: 4905–07.

Makova, K. D., and W. H. Li. 2002. "Strong male-driven evolution of DNA sequences in humans and apes." *Nature* 416: 624–626.

Malthus, T. R. 1798. *An Essay on the Principle of Population, as It Affects the Future Improvement of Society: With Remarks on the Speculations of Mr. Godwin, M. Condorset, and other Writers.* London: Printed for J. Johnson in St. Paul's Church-yard.

Mancuso, P., A. Gottschalk, S. M. Phare, M. Peters-Golden, N. W. Lukacs, and G. B. Huffnagle. 2002. "Leptin-deficient mice exhibit impaired host defense in Gram-negative pneumonia." *Journal of Immunology* 168: 4018–24.

Mann, D. R., M. A. Akinbami, K. G. Gould, and A. A. Ansari. 2000. "Seasonal variations in cytokine expression and cell-mediated immunity in male rhesus monkeys." *Cellular Immunology* 200: 105–115.

Margulis, S. W., J. Altmann, and C. Ober. 1993. "Sex-biased lactational duration in a human population and its reproductive costs." *Behavioral Ecology and Sociobiology* 32: 41–45.

Marin, P., S. Holmang, L. Jonsson, L. Sjostrom, H. Kvist, G. Holm, G.

Lindstedt, and P. Bjorntorp. 1992. "The effects of testosterone treatment on body composition and metabolism in middle-aged obese men." *International Journal of Obesity and Related Metabolic Disorders* 16: 991–997.

Marks, J. 1995. *Human Biodiversity: Genes, Race, and History.* New York: Aldine de Gruyter.

Marler, C. A., and M. C. Moore. 1988. "Evolutionary costs of aggression revealed by testosterone manipulations in free-living lizards." *Behavioral Ecology and Sociobiology* 23: 21–26.

Marlowe, F. 1999a. "Male care and mating effort among Hadza foragers." *Behavioral Ecology and Sociobiology* 46: 46–57.

—— 1999b. "Showoffs or providers? The parenting effort of Hadza men." *Evolution and Human Behavior* 20: 391–404.

—— 2000. "The patriarch hypothesis: An alternative explanation of menopause." *Human Nature* 11: 27–42.

Martens, J. W., S. Lumbroso, M. Verhoef-Post, V. Georget, A. Richter-Unruh, M. Szarras-Czapnik, T. E. Romer, H. G. Brunner, A. P. Themmen, and C. Sultan. 2002. "Mutant luteinizing hormone receptors in a compound heterozygous patient with complete Leydig cell hypoplasia: Abnormal processing causes signaling deficiency." *Journal of Clinical Endocrinology and Metabolism* 87: 2506–13.

Martin, C. W., R. A. Anderson, L. Cheng, P. C. Ho, Z. van der Spuy, K. B. Smith, A. F. Glasier, D. Everington, and D. T. Baird. 2000. "Potential impact of hormonal male contraception: Cross-cultural implications for development of novel preparations." *Human Reproduction* 15: 637–645.

Mattison, J. A., M. A. Lane, G. S. Roth, and D. K. Ingram. 2003. "Calorie restriction in rhesus monkeys." *Experimental Gerontology* 38: 35–46.

Mayr, E. 1997. "The objects of selection." *Proceedings of the National Academy of Sciences* 94: 2091–94.

———. 2004. *What Makes Biology Unique? Considerations on the Autonomy of a Scientific Discipline.* Cambridge: Cambridge University Press.

Mazer, N., D. Bell, J. Wu, J. Fischer, M. Cosgrove, and B. Eilers. 2005. "Comparison of the steady-state pharmacokinetics, metabolism, and variability of a transdermal testosterone patch versus a transdermal testosterone gel in hypogonadal men." *Journal of Sex Medicine* 2: 213–226.

Mazur, A. 1992. "Testosterone and chess competition." *Social Psychology Quarterly* 55: 70–77.

Mazur, A., and A. Booth. 1998. "Testosterone and dominance in men." *Behavioral and Brain Sciences* 21: 353–364.

Mendel, G., and J. Krizenecky. 1965. *Fundamenta Genetica: The Revised Edition of Mendel's Classic Paper with a Collection of 27 Original Papers Published during the Rediscovery Era.* Prague: Publishing House of the Czechoslovak Academy of Sciences; Moravian Museum.

Mieusset, R., and L. Bujan. 1995. "Testicular heating and its possible contributions to male infertility: A review." *International Journal of Andrology* 18: 169–184.

Miller, J. A. 1991. "Does brain size variability provide evidence of multiple species in *Homo habilis?*" *American Journal of Physical Anthropology* 84: 385–398.

Miller, J. M. 2000. "Craniofacial variation in *Homo habilis:* An analysis of the evidence for multiple species." *American Journal of Physical Anthropology* 112: 103–128.

Moggi-Cecchi, J. 2001. "Questions of growth." *Nature* 414: 595–597.

Møller, A. P., G. Sorci, and J. Erritzøe. 1998. "Sexual dimorphism in immune defense." *American Naturalist* 152: 605–619.

Moller, H. 1996. "Change in male:female ratio among newborn infants in Denmark." *Lancet* 348: 828–829.

Montanaro Gauci, M., T. F. Kruger, K. Coetzee, K. Smith, J. P. Van Der Merwe, and C. J. Lombard. 2001. "Stepwise regression analysis to study male and female factors impacting on pregnancy rate in an intrauterine insemination programme." *Andrologia* 33: 135–141.

Moore, S. L., and K. Wilson. 2002. "Parasites as a viability cost of sexual selection in natural populations of mammals." *Science* 297: 2015–18.

Morgan, T. H. 1915. *The Mechanism of Mendelian Heredity.* London: Constable.

Mrosovsky, N., and C. L. Yntema. 1980. "Temperature dependence of sexual differentiation in sea turtles: Implications for conservation practices." *Biological Conservation* 18: 271–280.

Muehlenbein, M. P. 2004. "Testosterone-mediated immune function: An energetic allocation mechanism in human and non-human primate males." Ph.D. diss., Yale University.

Muehlenbein, M. P., J. Alger, F. Cogswell, M. James, and D. Krogstad. 2005. "The reproductive endocrine response to Plasmodium vivax infection in Hondurans." *American Journal of Tropical Medicine and Hygiene* 73: 178–187.

Muehlenbein, M. P., and R. G. Bribiescas. 2005. "Testosterone-mediated immune functions and male life histories." *American Journal of Human Biology* 17: 527–558.

Murray, J., and G. Bertram. 1992. "The evolutionary significance of lifetime reproductive success." *The Auk* 109(1): 167–172.

Naeye, R. L., L. S. Burt, D. L. Wright, W. A. Blanc, and D. Tatter. 1971. "Neonatal mortality: The male disadvantage." *Pediatrics* 48: 902–906.

Nallella, K. P., R. K. Sharma, N. Aziz, and A. Agarwal. 2006. "Significance of sperm characteristics in the evaluation of male infertility." *Fertility and Sterility* 85: 629–634.

Nesse, R. M., and G. C. Williams. 1994. *Why We Get Sick: The New Science of Darwinian Medicine.* New York: Times Books.

Nicolich, M. J., W. W. Huebner, and A. R. Schnatter. 2000. "Influence of parental and biological factors on the male birth fraction in the United States: An analysis of birth certificate data from 1964 through 1988." *Fertility and Sterility* 73: 487–492.

Nicolosi, A., E. D. Moreira Jr., M. Shirai, M. I. Bin Mohd Tambi, and D. B. Glasser. 2003. "Epidemiology of erectile dysfunction in four countries: Cross-national study of the prevalence and correlates of erectile dysfunction." *Urology* 61: 201–206.

Nnatu, S. N., O. F. Giwa-Osagie, and E. E. Essien. 1991. "Effect of repeated semen ejaculation on sperm quality." *Clinical and Experimental Obstetrics and Gynecology* 18: 39–42.

Nunn, C. L., J. L. Gittleman, and J. Antonovics. 2000. "Promiscuity and the primate immune system." *Science* 290: 1168–70.

O'Carroll, R., C. Shapiro, and J. Bancroft. 1985. "Androgens, behaviour and nocturnal erection in hypogonadal men: The effects of varying the replacement dose." *Clinical Endocrinology* (Oxford) 23: 527–538.

Oberholzer, A., M. Keel, R. Zellweger, U. Steckholzer, O. Trentz, and W. Ertel. 2000. "Incidence of septic complications and multiple organ failure in severely injured patients is sex specific." *Journal of Trauma* 48: 932–937.

Odedina, F. T., J. O. Ogunbiyi, and F. A. Ukoli. 2006. "Roots of prostate cancer in African-American men." *Journal of the National Medical Association* 98: 539–543.

Offner, P. J., E. E. Moore, and W. L. Biffl. 1999. "Male gender is a risk factor for major infections after surgery." *Archives of Surgery* 134: 935–938; discussion 938–940.

Oldereid, N. B., J. O. Gordeladze, B. Kirkhus, and K. Purvis. 1984. "Human sperm characteristics during frequent ejaculation." *Journal of Reproduction and Fertility* 71: 135–140.

Olsen, G. W., K. M. Bodner, J. M. Ramlow, C. E. Ross, and L. I. Lipshultz. 1995. "Have sperm counts been reduced 50 percent in 50 years? A statistical model revisited." *Fertility and Sterility* 63: 887–893.

Orubuloye, I. O., and F. Oguntimehin. 1999. "Death is pre-ordained, it will come when it is due: Attitudes of men to death in the presence of AIDS in Nigeria." In J. Caldwell, P. Caldwell, J. Anarfi, K. Awusabo-Asare, J. Ntozi, I. O. Orubuloye, J. Marck, W. Cosford, R. Colombo, and E. Hollings, eds., *Resistances to Behavioural Change to Reduce HIV/AIDS Infection in Predominantly Heterosexual Epidemics in Third World Countries,* 101–111. Canberra: Health Transition Centre, National Centre for Epidemiology and Population Health, Australian National University.

Owens, I. P. 2002. "Ecology and evolution: Sex differences in mortality rate." *Science* 297: 2008–09.

Packer, C., M. Tatar, and A. Collins. 1998. "Reproductive cessation in female mammals." *Nature* 392: 807–811.

Page, K. C., C. M. Sottas, and M. P. Hardy. 2001. "Prenatal exposure to dexamethasone alters Leydig cell steroidogenic capacity in immature and adult rats." *Journal of Andrology* 22: 973–980.

Page, W. F., M. M. Braun, A. W. Partin, N. Caporaso, and P. Walsh. 1997. "Heredity and prostate cancer: A study of World War II veteran twins." *Prostate* 33: 240–245.

Paine, B. J., M. Makrides, and R. A. Gibson. 1999. "Duration of breast-feeding and Bayley's Mental Developmental Index at 1 year of age." *Journal of Paediatrics and Child Health* 35: 82–85.

Panter-Brick, C., D. S. Lotstein, and P. T. Ellison. 1993. "Seasonality of reproductive function and weight loss in rural Nepali women." *Human Reproduction* 8: 684–690.

Parazzini, F., C. La Vecchia, F. Levi, and S. Franceschi. 1998. "Trends in male:female ratio among newborn infants in 29 countries from five continents." *Human Reproduction* 13: 1394–96.

Partridge, L., and M. Farquhar. 1981. "Sexual activity reduces lifespan of male fruitflies." *Nature* 294: 580–582.

Patterson, N., D. J. Richter, S. Gnerre, E. S. Lander, and D. Reich. 2006. "Genetic evidence for complex speciation of humans and chimpanzees." *Nature Genetics* online at *www.nature.com*.

Peccei, J. S. 2001. "A critique of the grandmother hypotheses: Old and new." *American Journal of Human Biology* 13: 434–452.

Pelliniemi, L. J., K. Fröjdman, and J. Paranko. 1993. "Embryological and prenatal development and function of Sertoli cells." In L. D. Russell and M. D. Griswold, eds., *The Sertoli Cell,* 87–113. Clearwater, Fla.: Cache River Press.

Plas, E., P. Berger, M. Hermann, and H. Pfluger. 2000. "Effects of aging on male fertility?" *Experimental Gerontology* 35: 543–551.

Plavcan, J. M. 2000. "Inferring social behavior from sexual dimorphism in the fossil record." *Journal of Human Evolution* 39: 327–344.

—— 2001. "Sexual dimorphism in primate evolution." *American Journal of Physical Anthropology* Supplement 33: 25–53.

Plavcan, J. M., and C. P. van Schaik. 1997a. "Interpreting hominid behavior on the basis of sexual dimorphism." *Journal of Human Evolution* 32: 345–374.

—— 1997b. "Intrasexual competition and body weight dimorphism in anthropoid primates." *American Journal of Physical Anthropology* 103: 37–68.

Pope, H. G., Jr., E. M. Kouri, and J. I. Hudson. 2000. "Effects of supraphysiologic doses of testosterone on mood and aggression in normal men: A randomized controlled trial." *Archives of General Psychiatry* 57: 133–140.

Powell, I. J. 1997. "Prostate cancer and African-American men." *Oncology* 11: 599–605.

Quinn, M., and P. Babb. 2002. "Patterns and trends in prostate cancer incidence, survival, prevalence and mortality. Part I: International comparisons." *BJU International* 90: 162–173.

Raberg, L., M. Vestberg, D. Hasselquist, R. Holmdahl, E. Svensson, and J. A. Nilsson. 2002. "Basal metabolic rate and the evolution of the adaptive immune system." *Proceedings of the Royal Society of London, Series B* 269: 817–821.

Rasmussen, P. E., K. Erb, L. G. Westergaard, and S. B. Laursen. 1997. "No evidence for decreasing semen quality in four birth cohorts of 1,055 Danish men born between 1950 and 1970." *Fertility and Sterility* 68: 1059–64.

Reddy, S., M. Shapiro, R. Morton Jr., and O. W. Brawley. 2003. "Prostate cancer in black and white Americans." *Cancer Metastasis Review* 22: 83–86.

Reinisch, J. M., and S. A. Sanders. 1992. "Effects of prenatal exposure to diethylstilbestrol (DES) on hemispheric laterality and spatial ability in human males." *Hormones and Behavior* 26: 62–75.

Resnick, S. M., A. F. Goldszal, C. Davatzikos, S. Golski, M. A. Kraut, E. J. Metter, R. N. Bryan, and A. B. Zonderman. 2000. "One-year age changes in MRI brain volumes in older adults." *Cerebral Cortex* 10: 464–472.

Rhind, S. M., M. T. Rae, and A. N. Brooks. 2001. "Effects of nutrition and environmental factors on the fetal programming of the reproductive axis." *Reproduction* 122: 205–214.

Richerson, P. J., and R. Boyd. 2005. *Not by Genes Alone: How Culture Transformed Human Evolution.* Chicago: University of Chicago Press.

Ritz, E. 1985. "The clinical spectrum of hereditary nephritis." *Kidney International* 27: 83.

Robert, K. A., and M. B. Thompson. 2001. "Sex determination: Viviparous lizard selects sex of embryos." *Nature* 412: 698–699.

Roberts, A. C., R. D. McClure, R. I. Weiner, and G. A. Brooks. 1993. "Over-training affects male reproductive status." *Fertility and Sterility* 60: 686–692.

Röjdmark, S. 1987a. "Increased gonadotropin responsiveness to gonadotropin-releasing hormone during fasting in normal subjects." *Metabolism* 36: 21–26.

—— 1987b. "Influence of short-term fasting on the pituitary-testicular axis in normal men." *Hormone Research* 25: 140–146.

Rose, M. R. 1991. *Evolutionary Biology of Aging*. New York: Oxford University Press.

Ross, R., L. Bernstein, H. Judd, R. Hanisch, M. Pike, and B. Henderson. 1986. "Serum testosterone levels in healthy young black and white men." *Journal of the National Cancer Institute* 76: 45–48.

Ross, R. K., L. Bernstein, R. A. Lobo, H. Shimizu, F. Z. Stanczyk, M. C. Pike, and B. E. Henderson. 1992. "5-alpha-reductase activity and risk of prostate cancer among Japanese and US White and Black males." *Lancet* 339: 887–889.

Ruvolo, M., S. Zehr, M. von Dornum, D. Pan, B. Chang, and J. Lin. 1993. "Mitochondrial COII sequences and modern human origins." *Molecular Biology and Evolution* 10: 1115–35.

Ryan, M. J. 1980. "Female mate choice in a neotropical frog." *Science* 209: 523–525.

Safe, S. H. 1995. "Environmental and dietary estrogens and human health: Is there a problem?" *Environmental Health Perspectives* 103: 346–351.

Sahlins, M. D. 1976. *The Use and Abuse of Biology: An Anthropological Critique of Sociobiology*. Ann Arbor: University of Michigan Press.

Saidi, J. A., D. T. Chang, E. T. Goluboff, E. Bagiella, G. Olsen, and H. Fisch.

1999. "Declining sperm counts in the United States? A critical review." *Journal of Urology* 161: 460–462.

Salmimies, P., G. Kockott, K. M. Pirke, H. J. Vogt, and W. B. Schill. 1982. "Effects of testosterone replacement on sexual behavior in hypogonadal men." *Archives of Sexual Behavior* 11: 345–353.

Salonia, A., A. Briganti, P. Montorsi, T. Maga, F. Deho, G. Zanni, B. Mazzoccoli, N. Suardi, P. Rigatti, and F. Montorsi. 2005. "Safety and tolerability of oral erectile dysfunction treatments in the elderly." *Drugs and Aging* 22: 323–338.

Salzano, F. M., J. V. Neel, and D. Maybury-Lewis. 1967. "Further studies on the Xavante Indians. I. Demographic data on two additional villages: Genetic structure of the tribe." *American Journal of Human Genetics* 19: 463–489.

Sargis, E. J. 2002. "Paleontology: Primate origins nailed." *Science* 298: 1564–1565.

Schaal, B., R. E. Tremblay, R. Soussignan, and E. J. Susman. 1996. "Male testosterone linked to high social dominance but low physical aggression in early adolescence." *Journal of the American Academy of Child and Adolescent Psychiatry* 35: 1322–1330.

Schmid-Hempel, P. 2003. "Variation in immune defence as a question of evolutionary ecology." *Proceedings of the Royal Society of London, Series B* 270: 357–366.

Schröder, F. H. 1996. "Impact of ethnic, nutritional, and environmental factors on prostate cancer." In S. Bhasin, H. L. Gabelnick, J. M. Spieler, R. S. Swerdloff, C. Wang, and C. Kelly, eds., *Pharmacology, Biology, and Clinical Applications of Androgens*, 121–136. New York: Wiley-Liss.

Schroder, J., V. Kahlke, K. H. Staubach, P. Zabel, and F. Stuber. 1998. "Gender differences in human sepsis." *Archives of Surgery* 133: 1200–05.

Schurmeyer, T., and E. Nieschlag. 1982. "Salivary and serum testosterone under physiological and pharmacological conditions." In G. F. Read, D. Riad-Fahmy, R. F. Walker, and K. Griffiths, eds., *Immunoassays of Steroids in Saliva,* 203–209. Cardiff: Alpha Omega.

Sear, R., R. Mace, and I. A. McGregor. 2000. "Maternal grandmothers improve nutritional status and survival of children in rural Gambia." *Proceedings of the Royal Society B: Biological Sciences* 267: 1641–1647.

Seftel, A. D. 2006. "Male hypogonadism. Part I: Epidemiology of hypogonadism." *International Journal of Impotence Research* 18: 115–120.

Selevan, S. G., L. Borkovec, V. L. Slott, Z. Zudova, J. Rubes, D. P. Evenson, and S. D. Perreault. 2000. "Semen quality and reproductive health of young Czech men exposed to seasonal air pollution." *Environmental Health Perspectives* 108: 887–894.

Seo, J. T., K. H. Rha, Y. S. Park, and M. S. Lee. 2000. "Semen quality over a 10-year period in 22,249 men in Korea." *International Journal of Andrology* 23: 194–198.

Setchell, J. M., and A. F. Dixson. 2001. "Circannual changes in the secondary sexual adornments of semifree-ranging male and female mandrills *(Mandrillus sphinx)." American Journal of Primatology* 53: 109–121.

Shapiro, D. Y. 1992. "Plasticity of gonadal development and protandry in fishes." *Journal of Experimental Zoology* 261(2): 194–203.

Sherman, P. W. 1981. "Kinship, demography, and Belding's ground squirrel nepotism." *Behavioral Ecology and Sociobiology* 8: 251–259.

Shibata, A., and A. S. Whittemore. 1997. "Genetic predisposition to prostate cancer: Possible explanations for ethnic differences in risk." *Prostate* 32: 65–72.

Shimizu, H., R. K. Ross, L. Bernstein, R. Yatani, B. E. Henderson, and T. M. Mack. 1991. "Cancers of the prostate and breast among Japanese and white immigrants in Los Angeles County." *British Journal of Cancer* 63: 963–966.

Shipman, P. 1994. *The Evolution of Racism: Human Differences and the Use and Abuse of Science.* New York: Simon and Schuster.

Sholl, S. A., and K. L. K. R. 1990. "Androgen receptors are differentially distributed between right and left cerebral hemispheres of the fetal male rhesus monkey." *Brain Research* 516: 122–126.

Sinha, A., J. Madden, D. Ross-Degnan, S. Soumerai, and R. Platt. 2003. "Reduced risk of neonatal respiratory infections among breastfed girls but not boys." *Pediatrics* 112: e303.

Skarda, S. T., and M. R. Burge. 1998. "Prospective evaluation of risk factors for exercise-induced hypogonadism in male runners." *Western Journal of Medicine* 169: 9–12.

Smith, R. J., and S. R. Leigh. 1998. "Sexual dimorphism in primate neonatal body mass." *Journal of Human Evolution* 34: 173–201.

Sonino, N., C. Navarrini, C. Ruini, F. Fallo, M. Boscaro, and G. A. Fava. 2004. "Life events in the pathogenesis of hyperprolactinemia." *European Journal of Endocrinology* 151: 61–65.

Sorensen, M. V., and W. R. Leonard. 2001. "Neandertal energetics and foraging efficiency." *Journal of Human Evolution* 40: 483–495.

Spaas, P. G., and M. A. Bagshaw. 1990. "Prostate cancer occurring in identical twins: A case report." *Prostate* 16: 219–223.

Spratt, D. I., P. Cox, J. Orav, J. Moloney, and T. Bigos. 1993. "Reproductive axis suppression in acute illness is related to disease severity." *Journal of Clinical Endocrinology and Metabolism* 76: 1548–54.

Spratt, D. I., and W. Crowley Jr. 1988. "Pituitary and gonadal responsiveness is enhanced during GnRH-induced puberty." *American Journal of Physiology* 254: E652–657.

Spratt, D. I., L. S. O'Dea, D. Schoenfeld, J. Butler, P. N. Rao, and W. Crowley Jr. 1988. "Neuroendocrine-gonadal axis in men: Frequent sampling of LH, FSH, and testosterone." *American Journal of Physiology* 254: E658–666.

Sram, R. 1999. "Impact of air pollution on reproductive health." *Environmental Health Perspectives* 107: A542–543.

Stadtman, E. R. 2001. "Protein oxidation in aging and age-related diseases." *Annals of the New York Academy of Sciences* 928: 22–38.

Stanitski, D. F., P. J. Nietert, C. L. Stanitski, R. K. Nadjarian, and W. Barfield. 2000. "Relationship of factors affecting age of onset of independent ambulation." *Journal of Pediatric Orthopedics* 20: 686–688.

Stearns, S. C. 1989. "Trade-offs in life history evolution." *Functional Ecology* 3: 259–268.

—— 1992. *The Evolution of Life Histories.* Oxford: Oxford University Press.

Stein, D. G. 2001. "Brain damage, sex hormones and recovery: A new role for progesterone and estrogen?" *Trends in Neurosciences* 24: 386–391.

Steinmetz, H., L. Jancke, A. Kleinschmidt, G. Schlaug, J. Volkmann, and Y. Huang. 1992. "Sex but no hand difference in the isthmus of the corpus callosum." *Neurology* 42: 749–752.

Stevenson, D. K., J. Verter, A. A. Fanaroff, W. Oh, R. A. Ehrenkranz, S. Shankaran, E. F. Donovan, L. L. Wright, J. A. Lemons, J. E. Tyson, S. B. Korones, C. R. Bauer, B. J. Stoll, and L. A. Papile. 2000. "Sex differences in outcomes of very low birthweight infants: The newborn male disadvantage." *Archives of Disease in Childhood, Fetal and Neonatal Edition* 83: F182–185.

Suay, F., A. Salvador, E. Gonzalez-Bono, C. Sanchis, M. Martinez, S. Martinez-Sanchis, V. M. Simon, and J. B. Montoro. 1999. "Effects of competition and its outcome on serum testosterone, cortisol and prolactin." *Psychoneuroendocrinology* 24: 551–566.

Sullivan, E. V., A. Pfefferbaum, E. Adalsteinsson, G. E. Swan, and D. Carmelli. 2002. "Differential rates of regional brain change in callosal and ventricular size: A 4-year longitudinal MRI study of elderly men." *Cerebral Cortex* 12: 438–445.

Swaab, D. F., and E. Fliers. 1985. "A sexually dimorphic nucleus in the human brain." *Science* 228: 1112–15.

Tamimi, R. M., P. Lagiou, L. A. Mucci, C. C. Hsieh, H. O. Adami, and D. Trichopoulos. 2003. "Average energy intake among pregnant women carrying a boy compared with a girl." *British Medical Journal* 326: 1245–46.

Tanner, J. M. 1978. *Fetus into Man: Physical Growth from Conception to Maturity*. Cambridge, Mass.: Harvard University Press.

Tardif, S. D. 1997. "The bioenergetics of parental behavior and the evolution of alloparental care in marmosets and tamarins." In N. G. Solomon and J. A. French, eds., *Cooperative Breeding in Mammals*, 11–33. Cambridge: Cambridge University Press.

Tenover, J. S. 1992. "Effects of testosterone supplementation in the aging male." *Journal of Clinical Endocrinology and Metabolism* 75: 1092–98.

Thornhill, R. 1976. "Sexual selection and paternal investment in insects." *American Naturalist* 110: 153–163.

Thornton, J. W. 2001. "Evolution of vertebrate steroid receptors from an ancestral estrogen receptor by ligand exploitation and serial genome expansions." *Proceedings of the National Academy of Sciences* 98: 5671–5676.

Tomasson, R. F. 1984. "The components of the sex differential in mortality in industrialized populations, 1979–1981: Swedes, US whites, and US blacks." *Comparative Social Research* 7: 287–311.

Tortolero, I., G. Bellabarba Arata, R. Lozano, C. Bellabarba, I. Cruz, and J. A. Osuna. 1999. "Semen analysis in men from Merida, Venezuela, over a 15-year period." *Archives of Andrology* 42: 29–34.

Trivers, R. L., and D. E. Willard. 1973. "Natural selection of parental ability to vary the sex ratio of offspring." *Science* 179: 90–92.

Tsai, L. W., and R. M. Sapolsky. 1996. "Rapid stimulatory effects of testosterone upon myotubule metabolism and sugar transport, as assessed by silicon microphysiometry." *Aggressive Behavior* 22: 357–364.

Tuttle, M. D., and M. J. Ryan. 1981. "Bat predation and the evolution of frog vocalizations in the neotropics." *Science* 214: 677–678.

Uchida, A., R. G. Bribiescas, M. Kanamori, N. Hirose, J. Ando, and Y. Ohno. 2002. "Salivary testosterone levels in healthy 90 year old Japanese males: Implications for endocrine senescence." Paper presented at International Society for Human Ethology annual conference, Montreal.

Ulizzi, L., and L. Terrenato. 1987. "Natural selection associated with birth weight: V. The secular relaxation of the stabilizing component." *Annals of Human Genetics* 51: 205–210.

Umezaki, M., T. Yamauchi, and R. Ohtsuka. 2002. "Time allocation to subsistence activities among the Huli in rural and urban Papua New Guinea." *Journal of Biosocial Science* 34: 133–137.

Utami, S. S., B. Goossens, M. W. Bruford, J. R. de Ruiter, and J. A. R. A. M. van Hooff. 2002. "Male bimaturism and reproductive success in Sumatran orangutans." *Behavioral Ecology* 13: 643–652.

Valeggia, C. R., and P. T. Ellison. 2001. "Lactation, energetics, and postpartum fecundity." In P. T. Ellison, ed., *Reproductive Ecology and Human Evolution.* New York: Aldine de Gruyter.

van der Pal-de Bruin, K. M., S. P. Verloove-Vanhorick, and N. Roeleveld. 1997. "Change in male:female ratio among newborn babies in Netherlands." *Lancet* 349: 62.

Van Voorhies, W. A. 1992. "Production of sperm reduces nematode lifespan." *Nature* 360: 456–458.

Verthelyi, D. 2001. "Sex hormones as immunomodulators in health and disease." *International Immunopharmacology* 1: 983–993.

Visser, M. E., and C. M. Lessells. 2001. "The costs of egg production and incubation in great tits *(Parus major)." Proceedings of the Royal Society of London, Series B* 268: 1271–77.

Waites, G. M. 2003. "Development of methods of male contraception: Impact of the World Health Organization Task Force." *Fertility and Sterility* 80: 1–15.

Waldron, I. 1985. "What do we know about causes of sex differences in mortality? A review of the literature." *Population Bulletin of the United Nations,* 59–76.

Walker, A., and P. Shipman. 1996. *The Wisdom of the Bones: In Search of Human Origins.* New York: Knopf.

Walker, R., and K. R. Hill. 2003. "Modeling growth and senescence in physical performance among the Ache of eastern Paraguay." *American Journal of Human Biology* 15: 196–208.

Walker, R., K. R. Hill, H. Kaplan, and G. McMillan. 2002. "Age-dependency in hunting ability among the Ache of eastern Paraguay." *Journal of Human Evolution* 42: 639–657.

Wang, C., D. H. Catlin, B. Starcevic, A. Leung, E. DiStefano, G. Lucas, L. Hull, and R. S. Swerdloff. 2004a. "Testosterone metabolic clearance and production rates determined by stable isotope dilution/tandem mass spectrometry in normal men: Influence of ethnicity and age." *Journal of Clinical Endocrinology and Metabolism* 89: 2936–41.

Wang, C., G. Cunningham, A. Dobs, A. Iranmanesh, A. M. Matsumoto, P. J. Snyder, T. Weber, N. Berman, L. Hull, and R. S. Swerdloff. 2004b. "Long-term testosterone gel (AndroGel) treatment maintains beneficial effects on sexual function and mood, lean and fat mass, and bone mineral density in hypogonadal men." *Journal of Clinical Endocrinology and Metabolism* 89: 2085–98.

Wang, X., C. Chen, L. Wang, D. Chen, W. Guang, and J. French. 2003. "Conception, early pregnancy loss, and time to clinical pregnancy: A population-based prospective study." *Fertility and Sterility* 79: 577–584.

Wasser, S. K., and G. Norton. 1993. "Baboons adjust secondary sex ratio in response to predictors of sex-specific offspring survival." *Behavioral Ecology and Sociobiology* 32(4): 273–281.

Watts, D. P., J. C. Mitani, and H. M. Sherrow. 2002. "New cases of inter-community infanticide by male chimpanzees at Ngogo, Kibale National Park, Uganda." *Primates* 43: 263–270.

Waynforth, D., A. M. Hurtado, and K. R. Hill. 1998. "Environmentally contingent reproductive strategies in Mayan and Ache males." *Evolution and Human Behavior* 19: 369–385.

Wedekind, C. 1999. "Pathogen-driven sexual selection and the evolution of health." In S. C. Stearns, ed., *Evolution in Health and Disease,* 102–107. Oxford: Oxford University Press.

Welle, S., R. Jozefowicz, G. Forbes, and R. C. Griggs. 1992. "Effect of testosterone on metabolic rate and body composition in normal men and men with muscular dystrophy." *Journal of Clinical Endocrinology and Metabolism* 74: 332–335.

Westendorp, R. G., and T. B. Kirkwood. 1998. "Human longevity at the cost of reproductive success." *Nature* 396: 743–746.

Weston, G. C., M. L. Schlipalius, and B. J. Vollenhoven. 2002. "Migrant fathers and their attitudes to potential male hormonal contraceptives." *Contraception* 66: 351–355.

Whittemore, A. S., L. N. Kolonel, A. H. Wu, E. M. John, R. P. Gallagher, G. R. Howe, J. D. Burch, J. Hankin, D. M. Dreon, D. W. West, et al. 1995. "Prostate cancer in relation to diet, physical activity, and body size in blacks, whites, and Asians in the United States and Canada." *Journal of the National Cancer Institute* 87: 652–661.

Wiesenfeld, S. L. 1967. "Sickle-cell trait in human biological and cultural evolution: Development of agriculture causing increased malaria is bound to gene-pool changes causing malaria reduction." *Science* 157: 1134–40.

Wight, D. 1999. "Cultural factors in young heterosexual men's perception of HIV risk." *Sociology of Health and Illness* 21: 735–758.

Wilcox, A. J., C. R. Weinberg, J. F. O'Connor, D. D. Baird, J. P. Schlatterer, R. E. Canfield, E. G. Armstrong, and B. C. Nisula. 1988. "Incidence of early loss of pregnancy." *New England Journal of Medicine* 319: 189–194.

Williams, G. C. 1960. "Pleiotropy, natural selection, and the evolution of senescence." In B. L. Strehler, ed., *The Biology of Aging*, 332–337. Washington: American Institute of Biological Sciences.

——— 1991. "The dawn of Darwinian medicine." *Quarterly Review of Biology* 66: 1–22.

Wisniewski, A. B. 1998. "Sexually-dimorphic patterns of cortical asymmetry, and the role for sex steroid hormones in determining cortical patterns of lateralization." *Psychoneuroendocrinology* 23: 519–547.

Wolf, O. T., and C. Kirschbaum. 2002. "Endogenous estradiol and testosterone levels are associated with cognitive performance in older women and men." *Hormones and Behavior* 41: 259–266.

Wong, S. F., and L. C. Lo. 2001. "Sex selection in practice among Hong Kong Chinese." *Social Science and Medicine* 53: 393–397.

Wong, W. Y., C. M. Thomas, J. M. Merkus, G. A. Zielhuis, and R. P. Steegers-Theunissen. 2000. "Male factor subfertility: Possible causes and the impact of nutritional factors." *Fertility and Sterility* 73: 435–442.

Wong, W. Y., G. A. Zielhuis, C. M. Thomas, H. M. Merkus, and R. P. Steegers-Theunissen. 2003. "New evidence of the influence of exogenous and endogenous factors on sperm count in man." *European Journal of Obstetrics, Gynecology, and Reproductive Biology* 110: 49–54.

Wood, J. W. 1994. *Dynamics of Human Reproduction: Biology, Biometry, Demography.* New York: Aldine de Gruyter.

World Health Organization. 1999. *WHO Laboratory Manual for the Examination of Human Semen and Sperm–Cervical Mucus Interaction.* Cambridge: Cambridge University Press.

Worthman, C. M., and J. F. Stallings. 1994. "Measurement of gonadotropins in dried blood spots." *Clinical Chemistry* 40: 448–453.

Wrangham, R. W. 1999. "Evolution of coalitionary killing." *American Journal of Physical Anthropology* Supplement 29: 1–30.

Wrangham, R. W., J. H. Jones, G. Laden, D. Pilbeam, and N. Conklin-Brittain. 1999. "The raw and the stolen: Cooking and the ecology of human origins." *Current Anthropology* 40: 567–594.

Wrangham, R. W., and D. Peterson. 1996. *Demonic Males: Apes and the Origins of Human Violence.* Boston: Houghton Mifflin.

Wrangham, R. W., M. L. Wilson, and M. N. Muller. 2006. "Comparative rates of violence in chimpanzees and humans." *Primates* 47: 14–26.

Wright, A. L., C. J. Holberg, F. D. Martinez, W. J. Morgan, and L. M. Taussig. 1989. "Breast feeding and lower respiratory tract illness in the first year of life." *British Medical Journal* 299: 946–949.

Wu, F. C., T. M. Farley, A. Peregoudov, and G. M. Waites. 1996. "Effects of testosterone enanthate in normal men: Experience from a multicenter contraceptive efficacy study." *Fertility and Sterility* 65: 626–636.

Wynne-Edwards, V. C. 1962. *Animal Dispersion in Relation to Social Behaviour.* New York: Hafner.

Yaffe, K., L. Y. Lui, J. Zmuda, and J. Cauley. 2002. "Sex hormones and cognitive function in older men." *Journal of the American Geriatrics Society* 50: 707–712.

Yamanaka, M., and A. Ashworth. 2002. "Differential workloads of boys and girls in rural Nepal and their association with growth." *American Journal of Human Biology* 14: 356–363.

Yaqoob, M., H. Ferngren, F. Jalil, R. Nazir, and J. Karlberg. 1993. "Early child health in Lahore, Pakistan: XII. Milestones." *Acta Paediatrica* 82 Supplement 390: 151–157.

Zaldivar, M. E., R. Lizarralde, and S. Beckerman. 1991. "Unbiased sex ratios among the Bari: An evolutionary interpretation." *Human Ecology* 19: 469–498.

Zamudio, K. R., and B. Sinervo. 2000. "Polygyny, mate-guarding, and posthumous fertilization as alternative male mating strategies." *Proceedings of the National Academy of Sciences* 97: 14427–32.

Zhang, L., I. H. Shah, Y. Liu, and K. M. Vogelsong. 2006. "The acceptability of an injectable, once-a-month male contraceptive in China." *Contraception* 73: 548–553.

Zhang, Y., B. Nan, J. Yu, T. Snabboon, F. Andriani, and M. Marcelli. 2002. "From castration-induced apoptosis of prostatic epithelium to the use of apoptotic genes in the treatment of prostate cancer." *Annals of the New York Academy of Sciences* 963: 191–203.

Zihlman, A. L., and R. K. McFarland. 2000. "Body mass in lowland gorillas: A quantitative analysis." *American Journal of Physical Anthropology* 113: 61–78.

Zitzmann, M., and E. Nieschlag. 2001. "Testosterone levels in healthy men and the relation to behavioural and physical characteristics: Facts and constructs." *European Journal of Endocrinology* 144: 183–197.

Zumoff, B., L. K. Miller, and G. W. Strain. 2003. "Reversal of the hypogonadotropic hypogonadism of obese men by administration of the aromatase inhibitor testolactone." *Metabolism* 52: 1126–28.

ILLUSTRATION CREDITS

Figure 1   Chachugi. Photograph by Richard G. Bribiescas.

Figure 2   Monozygotic twins, one raised in a nurturing home, the other neglected. Reprinted by permission of the publisher from *Fetus into Man: Physical Growth from Conception to Maturity* by J. M. Tanner, Cambridge, Mass.: Harvard University Press, Copyright © 1978, 1989, by J. M. Tanner.

Figure 3   Mortality rates of wild chimpanzees, captive chimpanzees, and human hunter-gatherers (the Ache of Paraguay). Reprinted from *Journal of Human Evolution* 40, K. R. Hill, C. Boesch, J. Goodall, A. Pusey, J. Williams, and R. W. Wrangham, "Mortality rates among wild chimpanzees," 437–450, 2001, with permission from Elsevier.

Figure 4   The hypothalamic-pituitary-testicular axis. Drawing by Sam Ellison reprinted by permission of the publisher from *On Fertile Ground: A Natural History of Human Reproduction* by Peter T. Ellison, p. 254, Cambridge, Mass.: Harvard University Press, Copyright © 2001 by the President and Fellows of Harvard College.

Figure 5   Cumulative growth in healthy boys and girls. Reprinted by permission of the publisher from *Fetus into Man: Physical Growth from*

Conception to Maturity by J. M. Tanner, Cambridge, Mass.: Harvard University Press, Copyright © 1978, 1989, by J. M. Tanner.

Figure 6 Rates of growth in healthy boys and girls. Reprinted by permission of the publisher from *Fetus into Man: Physical Growth from Conception to Maturity* by J. M. Tanner, Cambridge, Mass.: Harvard University Press, Copyright © 1978, 1989, by J. M. Tanner.

Figure 7 A model of human male reproductive ecology. Reprinted from "Reproductive ecology and life history of the human male" by Richard G. Bribiescas, in *Yearbook of Physical Anthropology,* 2001.

Acknowledgments

This book can trace its roots to my initial field season among the Ache, a South American foraging population, in 1992. Further development was the result of countless conversations with colleagues and friends at professional conferences, cookouts, and hallway encounters. What I present in the book is a synthesis of what I consider to be important aspects of the evolution of the human male. This includes not only my own research but the ideas and results from many others who have been involved in research on human male life histories. I view my role as that of an organizer, attempting to compile and condense the vast body of research into a coherent account of what we know about human male life histories. Many people have contributed to the book, directly or indirectly. It's certain that I will overlook someone who deserves recognition. To those who remain unmentioned, I offer my apologies and sincere thanks.

I owe my male existence to wonderful parents, Asension and Belen. Thank you for tolerating my messy room so I could have more time to read. My sisters and brother, Dolores, Laura, and Chon, all share my love of learning and have been instrumental in my intellectual development throughout my life.

I owe a great deal, professionally and personally, to Peter Ellison, who mentored an unpolished but enthusiastic student from south central Los Angeles and gave him an opportunity to become a scientist. His example of professionalism and investment in his students left an indelible mark on me. I feel honored and fortunate to consider him a friend and colleague. At Yale University, I benefited from the support of Andrew Hill, Alison Richard, Eric Sargis, and David Watts. I have also been enriched by my graduate students, past and present, including Stephanie Anestis, Kate Clancy, Michael Muehlenbein, and Angélica Torres. Current members and alumni of the Reproductive Ecology Laboratory at Harvard University who have aided the development of my ideas include Gillian Bentley, Ben Campbell, Judith Flynn, Peter Gray, Carole Hooven, Grazyna Jasienska, Cheryl Knott, Susan Lipson, Matthew McIntyre, Mary O'Rourke, and Diana Sherry. Gillian Bentley provided enormously useful comments on early drafts.

I am also grateful to Irven DeVore and Richard Wrangham, who share my enthusiasm for understanding male life histories. DeVore's groundbreaking work among the !Kung San and his interest in understanding what makes men tick were central to the initiation of this book. Those not yet mentioned include William F. Crowley Jr., David Daegling, Terry Deacon, Seamus Decker, Donna Del Buco, Robert Dewar, Eduardo Fernandez-Duque, Sven Haakenson, Libra Hilde, Michal Jasienski, Jamie Jones, Frederika Kaestle, Greg Laden, Tim Laman, Bill Lukas, Sally McBrearty, Jonathan Padwe, Catherine Panter-Brick, Nadine Peacock, Rita Pizzi, Patrick Sluss, Mary Smith, Steve Stearns, Martha Tappen, Akiko Uchida, Claudia Valeggia, and Carol Worthman. In the field, I'm glad to have had the help and companionship of Richard Lawler.

To Kim Hill and A. Magdalena Hurtado, my thanks for giving me the opportunity to live and work with the Ache. Among the Ache, I have entire communities to thank, along with some special individuals. The residents of the colonias of Chupa Pou, Arroyo Bandera, and Puerto Barra all showed tremendous patience and goodwill during my investigations. To them I offer gratitude and good wishes. Individuals who were helpful in the conducting of my research included Angel Tatunambiangi, Carmen Bywangi, Blanco Bepagi, Ricardo Mirogi, Lorenzo Puaapiragi, Timoteo Turigi, Martin Achipurangi, and

Roberto Achipurangi. In Paraguay, Bjarne Fostervold and his wife Rosalba have been wonderful hosts. Their hospitality and companionship are most certainly treasured. I also thank Rosalino "Chalo" Cañete and his wife, Natividad, of Curuguaty, Paraguay, for their hospitality and friendship.

The Woodrow Wilson Foundation awarded me a fellowship that freed me from teaching responsibilities for a year. At the Wilson Foundation I would like to especially recognize Richard Hope, Bill Mitchell, and Sylvia Sheridan for their support. This book was also greatly improved by the comments of two reviewers.

At Harvard University Press, my warm thanks go to Michael Fisher for sharing my enthusiasm in the book's planning stages and for showing inexhaustible patience and support during its gestation. For Michael, in a spirit of apology, I offer a quotation from Douglas Adams, the author of *The Hitchhiker's Guide to the Galaxy:* "I love deadlines. I especially love the whooshing sound they make as they fly by." I am also indebted to Camille Smith of Harvard University Press for her editing expertise.

Finally, I thank my wife, Audrey. This paragraph has undergone numerous rewrites as I have attempted to do the impossible: to express the depth of my gratitude and love for her years of support and tolerance. Her unwavering advocacy and patience, while I toiled in front of the computer and this project monopolized many of our nights and weekends, is a debt I cannot repay. Without her, this book would not have been possible. Thank you dearly.

Index